Monster! Monster!

For my husband Robert—for his support throughout my monster mania, and for my sons Christopher, Kim and Phillip.

Monster! Monster!

A Survey of the North American Monster Scene

Betty Sanders Garner

ISBN 0-88839-357-1

Cataloging in Publication Data
Garner, Betty Sanders.
Monster! monster!
ISBN 0-88839-357-1

1. Monsters. I. Title.
QL89.G376 1995 001.9'44 C95-910639-1

Printed in Hong Kong

Cover Illustration: Kim Davis
Editing: Karen Kloeble
Production: Myron Shutty, Sandi Miller, Nancy Kerr

The author acknowledges with grateful thanks, Dr. Bernard Heuvelmans' permission to reproduce sketches of the types of sea serpents from his book, *In the Wake of the Sea-Serpents*.

Published simultaneously in Canada and the United States by

HANCOCK HOUSE PUBLISHERS LTD.
19313 Zero Avenue, Surrey, B.C. V4P 1M7
(604) 538-1114 Fax (604) 538-2262

HANCOCK HOUSE PUBLISHERS
1431 Harrison Avenue, Blaine, WA 98230-5005
(604) 538-1114 Fax (604) 538-2262

Contents

Preface

Between the covers of this book readers will discover over 100 "monsters"—all purported to live in North America. They have appeared in forest wildernesses, frolicked in lakes and rivers, and startled unsuspecting sailors at sea.

Those readers who find themselves having doubts should not be too hasty in their judgement. It is not fair to declare something a myth, or say it doesn't exist without first viewing all the evidence. It is also foolish to believe something doesn't exist merely because we don't yet have a specimen. New species are awaiting discovery today in the depths of the ocean, just as they are in remote areas of the earth.

Only since the turn of the century has science known the legendary mountain gorilla of Africa was indeed a real creature. Likewise, the giraffe's shortnecked relative, the Congo okapi, was not identified until 1900, and the pigmy hippo, in 1913. The Chinese giant panda was only a rumor until one was captured in 1936.

While some monster reports do tax the imagination, they leave a nagging question: do monsters exist? The reader will find his or her own answer.

Introduction

Foreword from The Great Sea-Serpent by A. C. Oudemans, Jzn., published by the author in 1892.

Voyagers and sportsmen conversant with photography are requested to take the instantaneous photograph of the animal: this alone will convince zoologists, while all their reports and pencil-drawings will be received with a shrug of the shoulders.

As these animals are very shy, it is not advisable to approach them with a steam boat.

The only manner to kill one instantly will be by means of explosive balls, or by harpoons laden with nitro-glycerine; but as it most probably will sink, when dead, like most of the Pinnipeds, the harpooning of it will probably be more successful.

If an individual is killed, take the following measurements:

1. Length of head from nose-tip to occiput.
2. Length of the neck from occiput to shoulders.
3. Length of the trunk from shoulders to tail-root.
4. Length of the tail from tail-root to end.
5. Distance from shoulders to fore-flappers.
6. Distance from shoulders to thickest part of the body.
7. Length of a fore-flapper.
8. Length of a hind-flapper.
9. Circumference of the head.
10. Circumference of the neck.
11. Circumference of the thickest part of the body.

12.Circumference of the tail-root.

Give a description of the animal, especially an accurate one of the head, the fore-flappers and the hind-flappers, and, if possible, make a sketch.

If but barely possible, preserve the whole skeleton, and the whole skin, but if this is utterly impracticable, keep the cleaned skull, the bones of one of the fore-flappers and those of one of the hind-flappers, four or five vertebrae of different parts of the backbone, neck and tail; and preserve the skin of the head, and a ribbon of about a foot breadth along the whole back of the neck, the trunk, and the tail.

Brachiosaurus

1

Echoes From The Past...

Long before the birth of man, in a time still largely shrouded in mystery, monsters walked the earth. They were the dinosaurs. They reigned as lords of the earth for 140 million years during the period of the earth's history known as the Mesozoic era. This period began about 225 million years ago and ended 65 million years ago.

The Mesozoic era is divided into three parts: the Triassic period, Jurassic period, and Cretaceous period. Dinosaurs first appeared about the last third of the Triassic period, but by the end of the Cretaceous, about 65 million years ago, they had all died out. But they left behind a wealth of evidence to their existence. People have been uncovering dinosaur bones for hundreds of years. Some of these fossils were over 215 million years old and were so large they gave rise to legends of giants stalking the earth.

Since the early 1800s, fossil hunters have found the bones of dinosaurs buried in the rocks of every continent except Antarctica (but they know dinosaur bones must exist there as well). Impressions of their skins, footprints in stone, even fossilized eggs and nests of eggs, have been found. Each new find has brought fresh information about how these fascinating beasts looked and lived.

In size, they ranged from as small as a chicken to monsters, perhaps as heavy as the blue whale—once thought to be the largest animal that ever lived. Some plodded about on all fours, while others walked and ran on their hind legs like ostriches.

Some had spikes, or crests, or frills. Some were protected by bony armor, but others had only their skin. The heads of some dinosaurs were very small; others, huge and surmounted by sharp horns. There were fierce carnivorous (flesh-eating) dinosaurs, and docile herbivorous (plant-eating) beasts. Some appear to have had feathers and others seem to have cared devotedly for their young.

The first recognized dinosaur remains in North America were found in New Jersey, in 1855. But the big push to discover dinosaurs did not begin until the late 1800s, when two wealthy rival American scientists, Othniel Charles Marsh and Edward Drinker Cope sparked the largest, longest quest for dinosaurs in the United States up to that time. Through their unceasing rivalry (they spied upon, brawled with and sabotaged each other through the 1870s), they were responsible for the most wide-ranging, systematic study of dinosaur life that existed until the middle of this century.

Between them, they opened up a vast unsuspected world of ancient life and added 136 species to the then already known nine dinosaur species. They brought us such colossal skeletons as Apatosaurus (once called Brontosaurus—because it supposedly made the ground shake as it walked); Barosaurus, Camarasaurus, Diplodocus, Allosaurus and Stegasaurus.

Since then, a veritable treasure trove of skeletons have been excavated from the Morrison sediments along the Como Bluff in Wyoming and an incredible tonnage of dinosaur remains from a quarry near Vernal, Utah in the Uinta Basin—now the Dinosaur National Monument. Texas, New Jersey, Kansas, Maryland, Massachusetts and New Mexico have also yielded significant finds. Some fine specimens can be seen at the Carnegie Museum in Pittsburgh and at the American Museum of Natural History in New York.

The first Canadian dinosaur remains were discovered in 1874 in the Morgan Creek district of Saskatchewan and the Milk River region near Comrey, Alberta. Beginning in 1897, valuable finds were also made in this area by the Geological Survey of Canada. These finds and others over the next three years finally climaxed in a very exciting period of history known as the Great Canadian Dinosaur Rush. From 1910 to 1917 the excitement and intense activity in western Canada, along the cliff-like banks of the Red Deer River, rivalled the famous Yukon Gold Rush although comparatively fewer people were involved. In the late 1970s and early 1980s, paleontologists found fossil beds in Dinosaur Provincial Park (a tract of barren hills and valleys straddling a section of Alberta's Red Deer River). These beds contained so many fossils that they could hardly move without stepping on bones of horned dinosaurs like Centrosaurus.

Deinonychus

New evidence has overturned many of the old beliefs about dinosaurs. They were not all, as previously thought, languid cold-blooded beasts; some, like the flesh-eaters, were warm-blooded, quick-moving and cared devotedly for their young. Scientists also no longer believe that dinosaurs were pea-brained giant reptiles. Most of them did have tough, reptilian-type skin, but many were more active than any reptiles living today. Their closest reptile ancestors gave rise to the crocodiles, yet their nearest living relatives are birds.

In 1972, bones of what seemed to be the monster of all prehistoric monsters were unearthed by Dr. James Jensen at Dry Mesa Quarry, in western Colorado. "Supersaurus," (super lizard) as it was nicknamed, outstripped all known North American Jurassic dinosaurs, including Brachiosaurus. With an estimated head height of fifty-four feet (16.5 m) and a total body length of eighty to one hundred feet (24 to 30 m), Supersaurus may have weighed as much as 100 tons (91 tonnes)—equal to a herd of fifteen African elephants.

Incredible as it seems, in 1979 an even larger specimen was found nearby. Tentatively named "Ultrasaurus," this giant beast may have been the largest dinosaur ever discovered, but both giants have yet to be scientifically named and studied. Ultrasaurus was estimated to be over 100 feet in length (30.5 m) and to weigh up to 150 tons (136 tonnes). It was a quarter larger than the largest mounted dinosaur and may have weighed as much as the blue whale!

Seventy million years ago dinosaurs seemed set to dominate the land forever. Then suddenly they all died out. What was the cause of this great dying? Many theories have been put forward as possible answers: their demise may have been the result of a sudden epidemic; a natural disaster like flood or volcanic eruption. They could have been killed by changing climatic conditions or some fluctuation in the sun's output of radiation. Or perhaps their end came about as the result of a single cataclysm such as a fiery encounter with a meteor.

Geologists have suggested a lump of rock (asteroid) about six to ten miles (10 to 16 km) across streaked in from space and hit the earth. The impact would have hurled dust and moisture up into the atmosphere darkening the sky everywhere for several years. Many plants and plant-eaters would have died. Perhaps the world froze briefly and then grew extremely hot; for although the dust settled, moisture still in the atmosphere created a greenhouse effect that stopped the sun's incoming heat from escaping. Overheating may have wiped out any dinosaur not already killed by cold or hunger.

No one theory, or combination of theories, can satisfactorily explain what caused the dinosaurs to become extinct and yet allowed crocodiles and other reptiles to survive. Perhaps we shall never know the true reason why these marvellous beasts, so long masters of the continents, disappeared completely from the earth. It will probably remain in the future, as it has in the past, one of the greatest mysteries to science.

Do monsters exist? We know they *did* exist—millions of years ago, when the world was young. We have the bones to prove it!

Plesiosaur

2

Myth Or Reality?

Since the dawn of history, people have been fascinated by tales of legendary monsters—huge hidden animals, known to individuals or local populations but not to the world at large. Strange creatures have walked the earth, swum in its seas and winged their way through its skies, striking terror and awe in their beholders.

Monsters have captured the imagination of poets, clerics, mariners and ordinary citizens, and have fired the ambitions of frauds and attention seekers. Yet these strange creatures have persistently avoided capture, seeming to secrete themselves in the world's most remote and mysterious regions. And there are still many such places for them to hide—even today, apart from the vast southern ice-cap, Antarctica, perhaps as much as one-tenth of the world is relatively unexplored. Remote lakes, inaccessible jungles, expanses of frozen tundra, hot deserts and immense sea areas remain almost unmapped or uncharted.

Monsters have figured prominently in the mythology of every people. Greek mythology described nine-headed water-serpents, flying horses, a beast that was half-lion and half-eagle, hairy giants who had one eye in the middle of their foreheads and a host of other weird creatures. Medieval sailors believed a great "Midguard" serpent called "Jormungard" encircled the world at the equator at the bottom of the sea. Ancient map makers depicted seas teeming with beasts of every description—spiny, winged, horned, sharp-toothed, many tentacled.

Sixteenth-century Bishop Olaus Magnus of Sweden, and later, Bishop Erick Pontoppidan of Norway, wrote extensively about the massive, many-armed sea monster, "Kraken" and the gigantic whale, "Leviathan." The latter was often mistaken for a small island by early sailors. Old stories tell of the Kraken so huge it could pluck a sailor from high in the rigging of a ship.

Both of these ancient monsters have now made the transition from mythical fancy to zoological fact. Modern science equates the Leviathan with the whale family. The Kraken is recognized as the giant squid which is found in many parts of the high seas. In our own hemisphere, they appear in the Gulf of Mexico, off the coast of Florida and they regularly come ashore along the coast of Newfoundland and Labrador.

Whereas today whales are perceived as benign, graceful giants of the deep, giant squids are the personification of our worst dreams—true monsters both in size and habit. They have two huge staring eyes, a parrot-like beak for tearing apart prey and ten long thin tentacles, two of which are much larger than the rest. Each tentacle is equipped with three rows of sucker-like discs that grip like a vice.

In 1871, and for almost ten years thereafter, giant squids, some as long as fifty-five feet (17 m), appeared off the coast of Newfoundland and Labrador. They crawled onto the beaches, became stranded and died. Ships made numerous sightings and there were reports of small boats being attacked.

One such attack occurred on October 26, 1873, while Daniel Squires, Theo Piccot and his twelve-year-old son, Tom were fishing for herring off the coast of Newfoundland. Noticing a large object in the water and thinking it to be a piece of wreckage, one of the men threw a boat hook to pull it in. Instantly, a giant creature with huge green eyes reared up its head and lunged at the gunwale with its beak. A huge arm and tentacle lashed itself around the boat and began to pull it down. Desperately, the fishermen flailed the water with the oars and bailed out their slowly capsizing boat. Tom grabbed a hatchet and began hacking at the tentacle. As he chopped through the grisly arm, the boat was freed from the hideous monster. Seizing the oars, the men rowed hard for shore. In their wake the giant squid slowly sank beneath the surface.

Tom sold his prize of the tough, leathery tentacle, measuring three and a half inches (9 cm) in diameter and nineteen feet

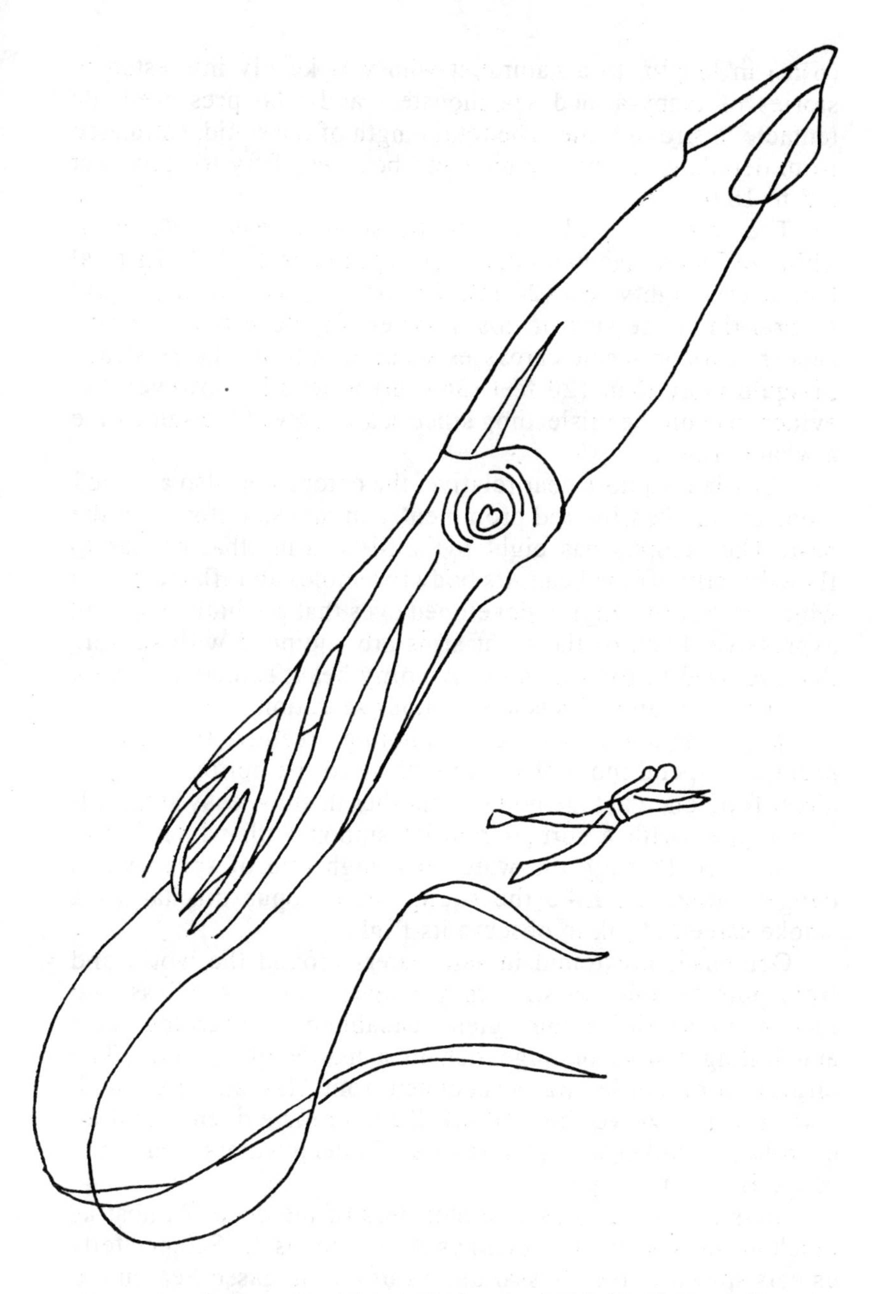

Comparison between the 80 foot (24.4 metres) Thimble Tickle squid and a human of average height.

(6 m) in length, to a naturalist who was keenly interested in stories of many-armed sea monsters and who preserved the tentacle in strong brine. The total length of the squid, estimated from details given by the men and boy, was fifty to sixty feet (15 to 18 m).

The largest squid accepted by science was captured at Thimble Tickle, Newfoundland on November 2, 1878. Its total length was eighty feet (24 m). The oceans may conceal squid several times the size of this monster. Eighteen inch (46 cm) sucker scars on whale carcasses seem to indicate the existence of squid more than 120 feet (36.6 m) in length. However, the evidence could be misleading since scars grow at the same time a whale grows.

The giant squid's near relative, the octopus, is also a "true" monster, and has figured prominently in monster stories in the past. The octopus has eight python-like arms that appear to flare directly off the head. Its body resembles an inflated bag in which are set two highly developed eyes that are both large and expressive. Each of the eight arms are equipped with suckers that are used to grab its prey. A horny beak, something like a parrot's, bites and injects a poisonous secretion.

While the squid is a free swimming creature, the octopus prefers to spend most of its time close to the bottom, moving about from one vantage point to another in slow, easy stages. It is equipped with a sort of ram-jet siphon with which it can propel itself through the water at a high rate of speed when danger threatens. Like the squid, the octopus discharges a smoke screen of ink to obscure its flight.

Octopuses are found in salt waters around the world and from pole to pole. In size, they range from as small as two inches (5 cm), to fearsome giants capable of overturning boats and killing a man or huge fish in a matter of seconds. The largest ever recorded was a specimen from Alaskan waters with a span of thirty-two feet (10 m). But there is evidence that an even larger, unknown octopus exists in deep waters near some of the Bahama Islands.

Eels also qualify as true monsters of the deep. Almost as much mystery surrounds them as the sea monster. Science tells us eels spawn in the abyssal depths of the Sargasso Sea, in the middle of the Atlantic Ocean. From there the thread-like elver swims upward and northward until it reaches the warm water of the Gulf Stream. Then it fans out west and east across North America and Europe. In North America they find their way through lakes and streams into the very center of the United States, up to the Great Lakes and well into Canada.

The conger and moray are the best known eels because of their size. The conger can grow to ten feet (3 m) and weigh hundreds of pounds (upwards of 45 kg). Congers have long, scaleless, snakelike bodies with long dorsal fins continuing around the tail as in other typical eels. They are bluish above and whitish below. Their wide mouths are armed with rows of closely spaced teeth. They are strong and voracious and often attack fish or animals of larger size.

According to a deep-sea salvage diver, the conger "is an ugly customer to tackle even when hauled aboard. Once it gets hold of a man's arm or leg, the only thing to do is cut off its head; even then its jaws have to be pried open."

The moray eel can grow to over twelve feet (3.6 m) in length and as big around as a telephone pole. It is vicious and has been known to attack swimmers and divers. It has a large mouth, filled with sharp teeth and will fight savagely to defend itself. There are many true stories of pearl divers who have been killed by it.

Thus, giant squid, octopus, conger and moray eels—particularly if overgrown or come upon unexpectedly, could all be likely ingredients for sea monster/sea serpent reports.

The earliest recorded account of a sea monster sighting was written by Aristotle in the fourth century. But, for hundreds of years now, seafarers and those who watch the sea from shore have been reporting the sightings of huge, unknown sea creatures. These reports frequently describe a creature strangely reminiscent of the ancient plesiosaur. In fact, long before Columbus, which was hundreds of years before man knew of the previous history of such beasts, scores of sworn statements by sea captains, bishops and other people of responsible position, described the plesiosaur in highly accurate detail. This huge saurian had a barrel-like body, an enormous long neck, a small head and four diminutive flippers, which resembled limbs. The plesiosaur and ichthyosaur dominated the oceans for many millions of years and probed every portion of the sea during the Mesozoic era.

Ichthyosaur

Although all the land dwelling dinosaurs became extinct about sixty million years ago, little is known about what went on in the sea. But the sea did preserve the coelacanth—a lobe-finned fish which first lived 300 million years ago. It had long been considered extinct at the time of its discovery in 1938, in the deep sea off the South African coast. If the coelacanth could survive from ages past, why not the plesiosaur or some of its contemporaries?

In modern times there are hundreds of seemingly reliable accounts of aquatic sightings and frequently the monster described is the plesiosaur. The most popular description of the famed Loch Ness monster of Scotland also gives a detailed word picture of this giant sea lizard.

With so many reliable reports of sightings of plesiosaur-type monsters, why does science not accept them as fact? Scientists are generally reluctant to say something exists unless there is proof of its existence. This could be either a living creature or a dead one. In the case of the coelacanth, the dead body was proof; thus, the scientific world is convinced that such creatures are still living. Unfortunately, the possibility of finding the carcass of a sea monster is very remote. If such a creature died at sea it would quickly sink to the bottom where it would be devoured by scavengers like the giant squid. Occasionally, huge dead monsters have been cast up on some forsaken beach and witnessed by a handful of non-scientific people. When local experts were called to examine the rotting remains, being at a loss for an explanation and wishing to protect their reputation, they have stated that the remains were those of a whale or basking shark. No scientist is willing to accept, without evidence, the fact that giant reptiles of the past have managed to transcend the bridge of time and remain undiscovered.

The logical thinker, however, can not help but wonder in light of the coelacanth's discovery. And we should not lose sight of the fact that science does not have all the answers. New species are awaiting discovery in the depths of the oceans just as they are in remote areas of the earth.

In recent years, photographs have sometimes accompanied sea serpent reports. Investigations of the accounts and the photographs have often led scientists to believe that most of the sea serpent/sea monster reports are simply cases of mistaken identity: masses of seaweed, waterlogged trees, or members of well-known living species—whales, basking sharks, giant squids, sea lions or manatees. There are, however, some descriptions and photographs of what appear to be living creatures that can not be explained or dismissed as man's imaginings, misconceptions, or the work of "hoaxers."

Until recently, it was thought that sightings of sea monsters were no longer common, the creatures being frightened away by noisy bulk carriers and supertankers. But this is not so, according to zoologist Bernard Heuvelmans, an avid collector of information on the subject from newspapers, magazines and personal correspondence. The number of reported sightings has remained relatively the same for the past 150 years.

Despite the lack of solid proof, belief in aquatic monsters today is widespread. For centuries, Scandinavians have taken their lake monsters for granted. "Necker" is believed to live in a lake near Stockholm and a monster with doglike head with great fins or ears at the back and short stumpy legs, in Lake Storsjon. "The Skrimsl," a forty-six foot long (14 m) monster lives in a lake in Iceland.

Russia has monsters in Lakes Vorota, Labynkr and Khaiyr. "The Piast," a kind of serpent generally believed to be a prehistoric animal about forty feet long (12 m), inhabits Lough Ree, and other lakes in Ireland. The pride of the British is "Nessie," the monster in Loch Ness, Scotland.

At least thirty lakes in Canada boast strange monsters lurking in their depths. The most famous of these is "Ogopogo,"

Coelacanth

who lives in Okanagan Lake, British Columbia, on the Pacific slopes of the Rockies.

The United States also has its share of aquatic monsters: "Slimy Slim," in Lake Payette, Idaho; several monsters in Bear Lake, Utah; and "Champ," resident monster of Lake Champlain, Vermont, to name only a few.

Monsters have always figured prominently in Indian and Inuit folklore. There are gods and demi-gods who inhabit underground worlds and the depths of the sea; werewolves and werebears. Stories of monsters in human form are common among most tribes.

To non-Indians, most of these monsters remain little more than legends. This is not so, however, in the case of a wild, hairy man-beast believed to inhabit the wilderness areas of the Pacific Northwest. He is known variously among Indian tribes as Saskehavas, Smy-a-likh, Seeahtik, and Sasquatch (wild-man). But the non-Indian has dubbed the creature "Bigfoot," because in its wake it leaves huge human-like footprints.

Present day Indians of the ancient eight nations of Totem-land, say that "the whiteman is hundreds of years late in catching on to the idea of Sasquatch." Perhaps so, but today there is much belief in this wild, giant man-beast who purportedly roams the vast forest lands of North America, from northern California to well north of Juneau. He is a living legend to thousands and a sworn reality to others. He has been frozen on film, tracked by hunters and studied by scientists and zoologists; yet, he still remains poised between myth and reality. Why do so many highly-educated and responsible people believe in the existence of monsters? And why do so many others want to believe? Perhaps in this great industrialized and highly mechanized world today, there is a need to know that a few creatures still hold their secrets close—live out their lives unharried and uncomplicated by mankind and its noisy pastimes. Perhaps they are remnants of the great dinosaurs!

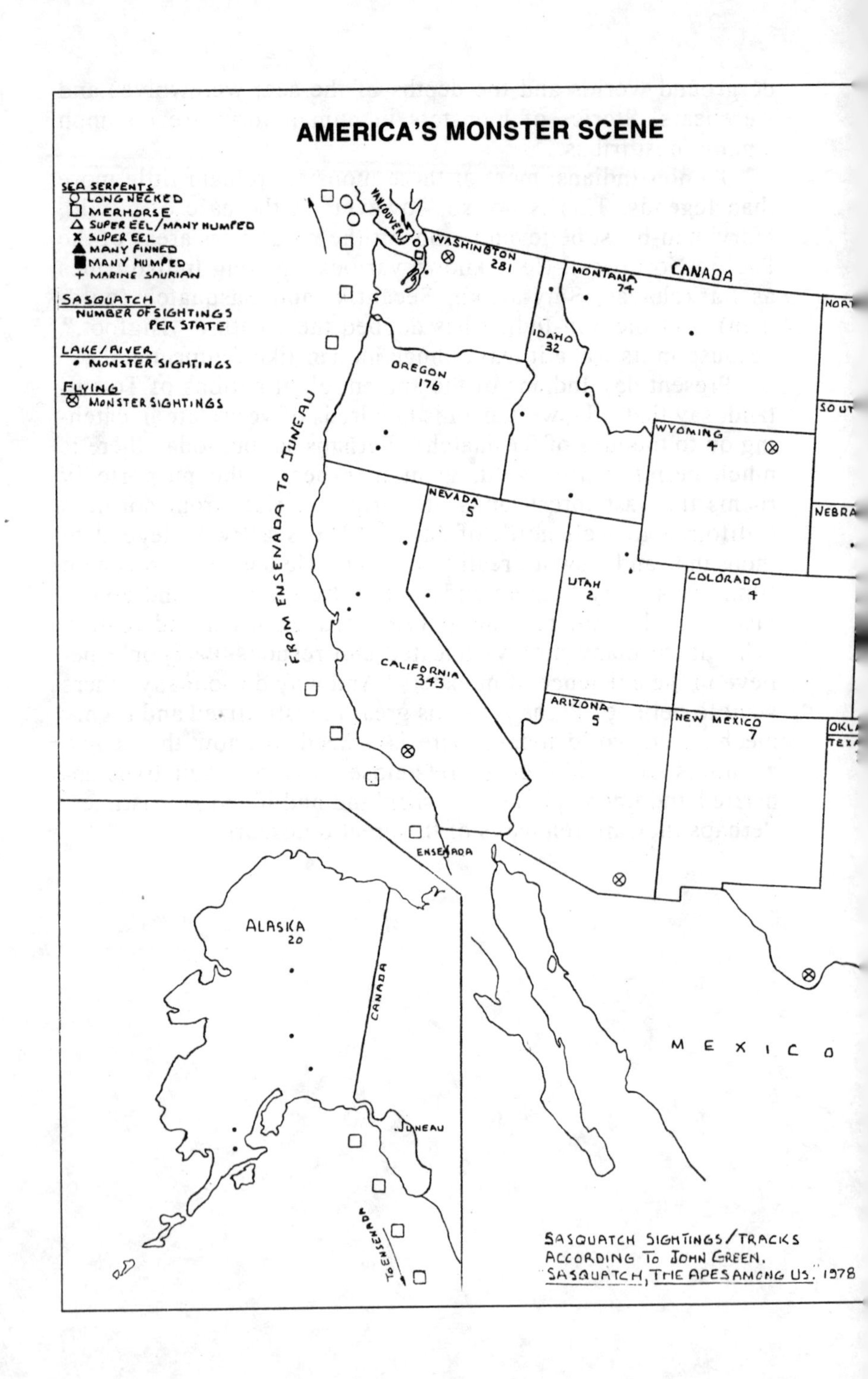
AMERICA'S MONSTER SCENE
SEA SERPENTS
LONG NECKED
MERHORSE
SUPER EEL/MANY HUMPED
SUPER EEL
MANY FINNED
MANY HUMPED
MARINE SAURIAN
SASQUATCH
NUMBER OF SIGHTINGS PER STATE
LAKE/RIVER
MONSTER SIGHTINGS
FLYING
MONSTER SIGHTINGS
VANCOUVER
WASHINGTON 281
MONTANA 74
CANADA
IDAHO 32
OREGON 176
WYOMING 4
NEVADA 5
UTAH 2
COLORADO 4
CALIFORNIA 343
ARIZONA 5
NEW MEXICO 7
FROM ENSENADA TO JUNEAU
ENSENADA
MEXICO
ALASKA 20
CANADA
JUNEAU
TO ENSENADA
SASQUATCH SIGHTINGS/TRACKS
ACCORDING TO JOHN GREEN.
SASQUATCH, THE APES AMONG US. 1978

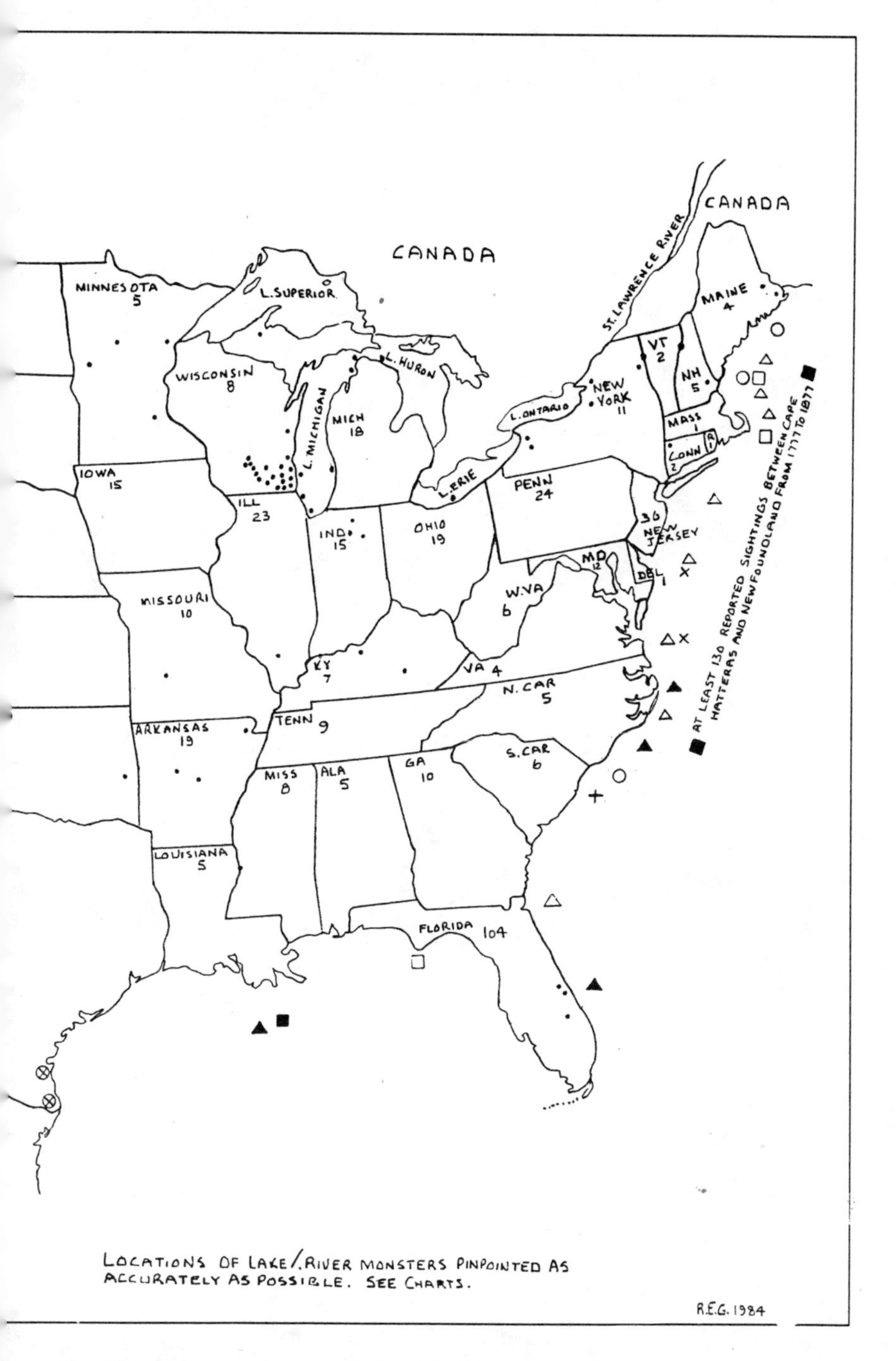

LOCATIONS OF LAKE/RIVER MONSTERS PINPOINTED AS ACCURATELY AS POSSIBLE. SEE CHARTS.

3

Bigfoot, The Elusive Wildman

You are alone in the wilderness, having made your way up a mountain trail to an old deserted mine. Here, you rest on a rock, savoring the crisp October air as you watch the comings and goings of little wild creatures on the forest floor. Then, your attention is drawn to movement in bushes on the other side of the clearing. At first you think it is a bear. Then, the bushes part and a huge creature, unlike anything you have ever seen before, steps into the open.

It is female, yet in size and stature, resembles a huge man completely covered with brown, silver-tipped hair. Its arms are thicker than a man's, very long and reach almost to its knees. Its head seems small compared to the rest of its body and sits on its shoulders with no neck. Its eyes are small and black, like a bear's. As it walks, it places the heel of its foot down first, exposing the gray-brown soles of its feet. It comes to within twenty feet (6 m) of where you are sitting, then squatting on its haunches, it begins to strip leaves and berries from the bushes with its teeth, which are white and even. Reason tells you there can't be such a creature! It must be an actor dressed up to look half-animal and half-man. Then you realize it would be impossible to fake such a specimen.

Suddenly, the creature notices you watching. The look of amazement on its face is so comical you can hardly keep from laughing. Still in a crouched position, it backs up three or four short steps, then abruptly straightens up and hurries back across the clearing. Before disappearing into the bushes, it turns and looks back.

Later, you make enquiries and find there was no film company in the area at the time. You realise you have had a real-life encounter with the elusive monster, Bigfoot! So it was for a man named William Roe, who experienced this in 1955, on Mica Mountain, near Tete Jaune Cache, British Columbia.

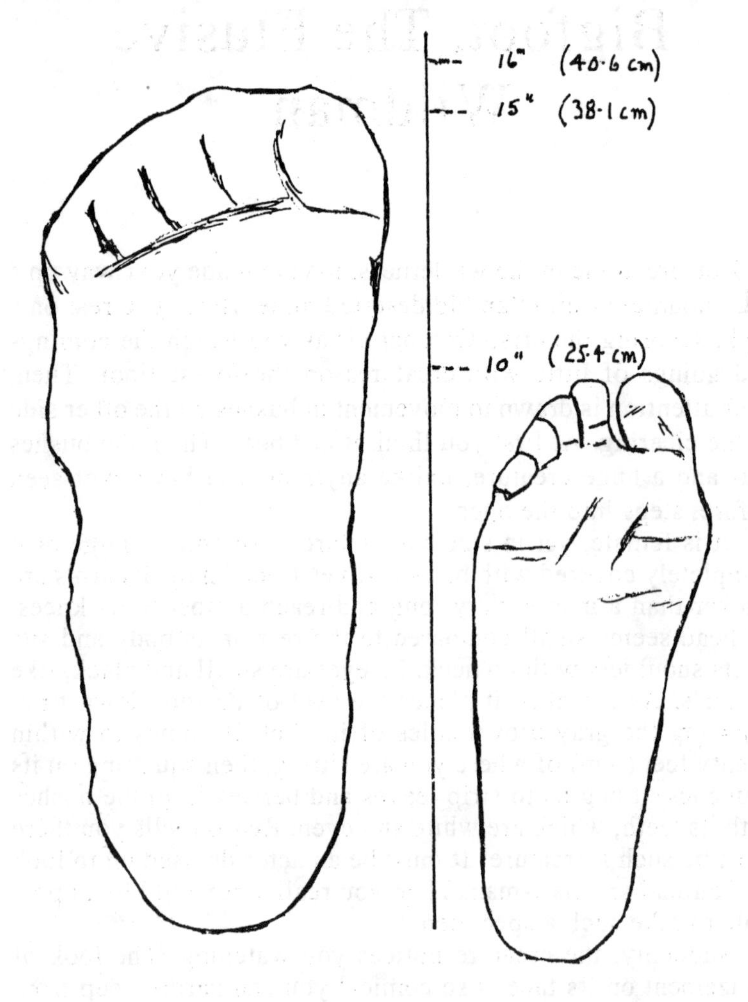

Comparison of an alleged Sasquatch foot (left) and a human foot.

People have been telling of sightings and similar encounters with this incredible wildman, for several hundred years. Today, there are at least four sightings every year. He is generally described as half-man, half-beast, six to nine feet (1.8 to 2.7 m) tall and weighing up to 1,000 pounds (453.6 kg). He is entirely covered with reddish-brown or black hair, except for his hands and face. His head is ape-like with low brow and sloping forehead rising to a crest at the back of the skull. His eyes glow like many smaller nocturnal animals such as cats, porcupines or skunks, when caught suddenly in the glare of artificial light. He has long, powerful arms, a heavy torso and is stoop-shouldered. Walking upright like a man, he takes strides of six to eight feet (1.8 to 2.4 m) and leaves deep, human-like foot-prints of up to twenty-one inches (53.3 cm) long. His home is believed to be in the Pacific Northwest where vegetation is extremely thick and high and where primitive conditions still prevail.

Since the early 1800s, sightings of this wildman have been reported from Alaska to California; in Mexico and Guatemala; and even into the Andes of South America. Similar creatures are also purported to live in the Himalayas, where they are called "Yeti," or "Abominable Snowmen." In the Caucasus region of Russia, they are known as "Almas," or "forestmen."

Although descriptions of the man-beast vary slightly, it is possible that they are all part of the same general species. Scientists have suggested that both the yeti and sasquatch are descendants of a prehistoric giant ape known scientifically as Gigantopithecus. Fossilized giant jawbones and teeth of these creatures have been found in the Siwalk Hills of northern India, and in Liv-Cheng County, Kwangsi province, southern China. These hairy, giant human-like creatures could have migrated to North America from their Asian mountain habitat at various times during the Ice Age, when the land bridges existed.

Before the early 1800s, a great number of sightings and even face-to-face encounters went unrecorded. Witnesses were afraid to tell about sightings lest they be subjected to ridicule and cries of "hoax."

Such was the case with Albert Ostman, a prospector who was kidnapped by a sasquatch on the British Columbian coast, in 1924. His story seemed so bizarre at the time, he kept it to himself thinking no one would believe him.

At the time of the incident, Ostman had set up camp near Toba Inlet. Food was stolen from his campsite the two previous nights, but he stayed on as it was a particularly good location, near a stream. The following night, he was shaken awake by something carrying him. After what seemed like hours, he was finally dumped on the ground. When he crawled out of his sleeping bag, he found himself face to face with a sasquatch family. The female seemed upset that the "old man" had brought home such a creature.

Ostman said they did not try to harm him, but seemed more curious about him than anything else. He was unable to escape for six days. He finally managed this when he offered the old male sasquatch some snuff. The creature swallowed the entire contents of the box, licking it clean. He immediately became ill and in the confusion that followed, Ostman was able to make his escape. Many years passed before Ostman finally told his story and then only after hearing of other people's experiences.

Ostman made a point of saying that the creatures did not harm him during his time with them, but they "watched curiously" as he cooked his food. This aspect has been mentioned in other sightings. As a rule, the sasquatch gives man a wide berth, but now and again his curiosity gets the better of him. It is at such times that farmers find footprints on their property. In some cases the creatures appear to have made a complete tour of inspection around the house, barn and other buildings. Sometimes they have appeared at windows, their eyes glowing, amber, red or green in the night, but they have always disappeared immediately if anyone went outside.

One old and interesting report harks back to 1884. The July 4 edition of *The Daily British Colonist*, Victoria, British Columbia, told of the capture of a boy sasquatch. He was found by railroad men, on the regular Lytton to Yale line, lying beside the track. Apparently he had fallen from the steep bluffs, high above. When the crew stopped the train, the young male sasquatch sprang up, uttering a sharp quick bark, and began climbing the steep bluff. The railroad men gave chase and succeeded in capturing him.

"Jacko," as he was named by his captors, was described as being "half-man and half-beast." He was approximately four feet seven inches (1.4 m) tall and weighed 127 pounds (57.7 kg). He had long black, strong hair and resembled a human being except his body, excluding hands and feet, was covered with glossy one inch (2.5 cm) long hair. His forearm was much longer than a man's and he was extraordinarily strong. He could break a stick by wrenching and twisting it in a way no human could.

Jacko was taken to Yale and exhibited to the public. His favorite food was berries and he enjoyed fresh milk.

Since the old, yellowed newspaper clipping of this report has now disappeared, there is no way of verifying the story, or of knowing what eventually happened to Jacko. Hopefully, he was released and allowed to return to his own kind.

In the fall fishing season of 1967, Marietta, Washington, experienced a rash of sightings. A dark, upright creature estimated to be twice the size of a man was seen on several occasions sitting on a tree stump beside the Nooksack River, northwest of the San Juan Archipelago. On other occasions it was seen wading on the tide flats. One fisherman told of an incident while he was night fishing.

Feeling a tug at the end of his net, he discovered a dripping wet sasquatch standing in the water pulling the net towards him, no doubt bent upon helping himself to the fisherman's catch! The man's shouts brought other fishermen down the river. They shone their spotlights on the creature as it stood in the water, pulling on the net. It promptly dropped the net and disappeared into the night.

In another incident, a woman who was also night fishing received a nasty shock when a sasquatch suddenly stood up in the water some four feet (1.2 m) or so from the side of her boat. Next morning, tracks, thirteen and a half inches (34 cm) long and with "no discernible arch," were found in the mud leading out of the water and returning again.

These sightings indicated that, not only was Bigfoot quite at home in the water, he was also an intelligent creature—he knew how to remove the fish from the net.

A strange story was reported from the State of New Jersey, in 1977. Early on the morning of May 12, Mrs. Barbara Sites, who lived with her husband and six children on a 120 acre (48.5 hectares) farm on Wolfpit Road, Wantage Township, went to put the heifers out to pasture. For some reason they laid their ears forward and stubbornly refused to leave the barn. From the nearby swamp she heard a noise like a woman screaming.

When Mrs. Sites went to check the rabbits, she discovered the door to the shed where they were kept had been ripped off its frame. Inside she found the rabbits all dead and their crushed bodies neatly lined up on top of their cages. That night, a big hairy creature, with "no neck," red eyes and about eight feet (2.4 m) tall, was seen standing by the corner of the shed. The Sites' dog, an animal of some seventy pounds (31.75 kg) in weight, went after the creature. With a sweep of its huge arm it threw the dog twenty feet (6 m) away. The dog promptly took to its heels and didn't return until the next day. Mrs. Sites was so upset by the experience that she took her children to her mother's home in Hamburg.

That night, the creature appeared again at the Sites' farm, this time in the old chicken coop. It could be seen clearly by the dawn-to-dusk light, which was on. The huge animal came out of the coop and stood under a nearby tree. Mr. Sites, his brother-in-law and a friend began shooting at the creature, which then turned and ran up the shoulder of the road in the tall grass. Mr. Sites gave chase in his pick-up truck, but the huge figure ran into the fields and disappeared. It returned to the site several times after that and was seen by other people walking through the fields, apparently carrying something white under its arm.

Autopsies performed on the rabbits revealed their heads had been torn off and the skulls and hindquarters crushed. Three of the torsos revealed broken ribs and internal organs burst from pressure. There were teeth marks but no puncture wounds. A dentist's report, after study of the marks, indicated they were made by a creature with teeth and jaws twice as large as those of an adult human being.

The results of the autopsies certainly supported the Sites' claim of a Bigfoot presence on the farm. The incident also brought to mind an interview journalist/investigator John Green had with a man in the fall of 1969. Green published the full text in his book, *On The Track of the Sasquatch*. Apparently the man chanced upon a Bigfoot family on a mountain trail in Oregon. The male, female and baby were engaged in turning over a pile of loose rocks, sniffing each one and then carefully stacking them in piles. The man soon realized they were searching for nests of hibernating rodents.

When the male got the scent of a nest "he really made the

rocks fly." He lifted out the grass nest which contained six to eight rodents, then taking a rodent in hand, the sasquatches proceeded to eat them as we would eat a banana.

This information and that in the Sites' report, suggested the sasquatch diet might also include small animals. Rodents, deer-mice gathered from the forest floor and voles from bogs and Alpine meadows, could all provide valuable protein for the creature's diet.

Up until 1967, Bigfoot was still largely considered a myth, despite the hundreds of reported sightings and sworn testimonies from witnesses. Now, suddenly in October that year, there came startling news that Roger Patterson of Yakima, Washington had actually captured Bigfoot on film! Apparently, Patterson and Bob Gimlin, also of Yakima, were making daily patrols on horseback in the Bluff Creek Valley, in northern California, looking for tracks on the sand bars along the creek. Not only did they find footprints, they also chanced upon a female sasquatch. Patterson's horse reared in fright at the sight of the creature, throwing him to the ground. The first few frames of the movie are very shaky as Patterson endeavored to regain his feet and follow the sasquatch, filming as he went. When the film settles down, for a few seconds a huge, heavily built creature with glistening dark hair, looks directly into the camera. Then it turns and swinging its heavy arms, crosses open ground in smooth loping strides and eventually disappears into the forest.

The film did not settle the question of the sasquatch's existence. Instead, it has been subjected over the years to much scrutiny and discussion among scientists and investigators in general. Some have declared it an out and out hoax; a man dressed up in a "monkey-suit," or some sort of "machine thing." But so far no scientist has been able to prove that it was anything other than what Patterson said it was—a living record of an encounter with a real sasquatch.

In May, 1978 the film was shown at the First Annual International Conference on Humanoid Monsters, which was held at the University of British Columbia. It was viewed by an impressive gathering of scientists, scholars and sasquatch hunters and investigators. Among the latter: John Green, Rene Dahinden, Dennis Gates and Peter Byrne, the former big game hunter who once searched for the yeti in the Himalayas. In the opinion of Dmitri Bayonov, of the Darwin Museum of Moscow and Russian biomechanics expert Dmitri Donskoy, the star of the movie was indeed a genuine sasquatch. The creature is neither ape nor man but passes tests "for distinctiveness and consistency and naturalness." The motion of the creature's legs and arms resembles that of a cross-country skier. Being a flat footed animal, the sasquatch leans forward as it walks and absorbs shock by walking with its knees bent.

From its footprints, the Russians conclude that the creature's feet are straight and that it digs its toes into the ground to propel itself forward. In their opinion, it would be impossible for a hoaxer to reproduce such a gait.

Dr. M. Jeanne Koffman, of the Darwin Museum in Moscow, who's paper was presented in absentia at the conference, likened the sasquatch to the forestmen, or Almas of Russia, that inhabit the mountainous Caucasus region—the isthmus between the Caspian and Black Seas. During her twenty years in the field, Dr. Koffman observed a group of five of these "relic humanoids," in the mountains of the southern Russian Steppes. She indicated the similarity extended even to the bad smell associated with the males.

Now rarely seen, a generation or so ago the alma was part of the landscape. People offered them food and even clothes. The adults are much like the sasquatch, but are smaller in size. They feed on fruits, berries, a variety of wild and cultivated plants, small animals, bird's eggs and foods such as dairy products, meat, honey and porridge, given them by man. Dr. Koffman also said, there has been a catastrophic drop in their numbers over the past forty to fifty years. Sightings are becoming more rare and their population is now at a "critical level."

Despite the fact no bones or carcasses have been found to substantiate their existence, the alma is an accepted fact in Russia. They are believed to bury their dead.

Where the sasquatch is concerned, the people of North America are harder to convince. There are too many questions without satisfactory answers. How could this huge, bipedal creature exist in this day and age, without our knowing? How could it stay so well hidden? Wouldn't we find some remains of its dead? And if Bigfoot were real, wouldn't scientists find some evidence of its existence?

Again we must remind ourselves, science does not have all

the answers, and it is foolish to believe something does not exist merely because we don't yet have a specimen. New species are awaiting discovery today, in remote areas of the earth, just as they are in the depths of the oceans. Only since the turn of the century has science known the legendary mountain gorilla of Africa was indeed a real creature. Likewise, the giraffe's short-necked relative, the Congo okapi, was not identified until 1900 and the pigmy hippo, in 1913. The Chinese giant panda was also unknown until one was captured in 1936. People had told of seeing all these creatures long before scientists proved their existence.

Science demands more tangible evidence than mere verbal reports before it can declare something exists. So far, all we have are the tantalizing glimpses and odd encounters Bigfoot allows us, together with thousands of huge, human-like footprints he leaves on mountain roads, creek banks and in the snow. Organic debris or droppings, unlike any deposited by North American game animals, have been found. Also, coarse long dark hair has been seen clinging to tree branches. But nothing as convincing as the teeth or jawbones of Gigantopithecus found in northern India and southern China.

Dr. Grover Krantz, associate professor of anthropology at Washington State University, thinks the reason nobody has ever found the corpse or skeleton of a sasquatch is the same reason bear hunters never find bears that have died natural deaths, "unless he's killed, an animal will hide himself before he dies. Finding a skeleton would be extraordinary."

Sightings of this mysterious wildman now number over 1,500 and footprints in the tens of thousands. Of these footprints, some are obviously fakes. But not all of them. In many cases the tracks made are too complex to be the work of hoaxers. The length of stride, for instance, far exceeds the capability of man, and the depths of imprints, up to three inches (7.6 cm), would require additional weight of several hundred pounds.

People have seen Bigfoot in the Cascade Mountains of British Columbia and Washington; Oregon and northern California; the Everglades of Florida; the Blue Ridge Mountains of Arkansas and the bogs of Louisiana. They have seen him standing by roadsides, swimming in lakes, wading in rivers, eating berries from bushes and watching them from the edge of the woods. But at no time has his appearance proposed a threat to people.

As in the case of William Roe's face-to-face encounter with the sasquatch on Mica Mountain, near Tete Jaune Cache, B.C., the creature has been just as surprised as they by the encounter.

In recent years, sightings have been most heavily concentrated in Florida and Texas and less heavily in the Midwest, especially Illinois, Arkansas, Missouri, Iowa, Indiana and Michigan.

Most sightings take place in summer and fall. This may be the result of man's increased outdoor activity during those seasons. The small number of winter reports suggests these creatures sleep through the cold weather—a conclusion reached in Russia, with the yeti. Or it could be that in winter Bigfoot leaves the high country for the wild, uninhabited beaches of northern California, where exceptional low winter tides make clam hunting and scavenging of tide pools possible.

In keeping with his largely nocturnal habits he could sleep in the forest by day and hunt the beaches by night; the incoming tide washing away all trace of his footprints. With the passing of winter he could again be called to the mountain wilderness, with its promise of fresh abundance of young plants, berries and small animals.

In recent years searches by non-professionals for this elusive giant have become very numerous. Armed with tranquilizer guns, hunters have set baited traps, set out high pitched bell-toned lures and have strung the wilderness with electronic trip wires. They stalk him with high-powered rifles and search lights mounted on trucks probe the forest darkness for his hideaways. Even dogs are being used to track him down; for the race to be the first to bring in a sasquatch, dead or alive, is very great and the financial rewards could be enormous.

If indeed Bigfoot exists, it is right we should know about it. Then, he should be left to live his life his own way, in peace. There are a few dedicated hunters in the field, like John Green, Rene Dahinden and Peter Byrne, who have spent more than forty years in the search. Perhaps by the efforts of these dedicated individuals we shall one day have final proof—or, perhaps Bigfoot will reveal himself in his own way, as a real, living creature who also has his place in the scheme of things.

Pterosaur

4

The Flying Phantom Of The Old West

Sometimes monster stories seem so bizarre as to defy all reason, yet, coupled with other factual events, they somehow take on flesh. Such is the case with reports of the Flying Phantom of the Old West.

From time to time in the United States, there are reports of horses and cattle mysteriously disappearing without a trace from corrals. Police departments, called in to investigate, have been unable to solve the mystery. It seems almost as though some giant airborne predator swooped down and carried the animals cleanly away.

Helicopters, man made mechanical birds of modern times, could do this job very nicely and no doubt are responsible for some of the disappearances. But many of the reports occurred in the 1800s, long before helicopters were invented—so they could not have been a possible cause at that time.

No flying creature large enough or strong enough to carry away a cow exists in modern times. But they did exist in the days of the dinosaurs: pteranodon, or quetzalcoatlus could have performed this feat very nicely.

Pteranodon of Kansas, possibly the ugliest and one of the largest airborne creatures of all time, had a wingspan of up to twenty-seven feet (8 m). Its body was as big as a turkey's and was very light in proportion to its large wing size. It weighed about sixty pounds (27 kg) and was a graceful, highly intelligent creature. Like other pterosaurs, or flying reptiles, its leathery wings were featherless membranes stretching from the extended fourth "fingers" to the hind legs. The other fingers were tipped with claws that probably allowed it to anchor to cliffs or tree limbs. Its long beak was toothless; designed for scooping up fish from the sea.

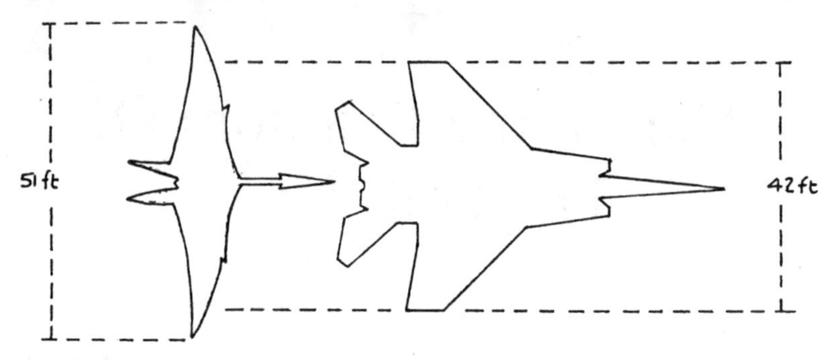

The 50 foot wingspan of the pterosaur compared to a jet fighter.

Some pteranodons had a crest of bone, resembling the crest of a blue jay, behind the head. This was possibly a feature of the male. Pteranodon lived like a reptilian albatross, sometimes spending days away from the cliffs that were its home, soaring over the extensive Cretaceous seas and plummeting down at intervals to snatch up a fish or small reptile into its cruel beak. Fossils suggest it sometimes crash landed in the sea.

Quetzalcoatlus, the largest flying creature known to have existed, had a gigantic wingspan of fifty-one feet (15.5 m) or more—larger than that of many airplanes. Its wings were so huge, it probably never flapped them. To take off, it may have jumped from high places and then glided through the air. Its ability to soar was probably akin to that of today's vultures.

Like pteranodon, and other pterodactyls, quetzalcoatlus' body was covered with hairless wrinkled skin. It was a scavenger—the greatest of them all! Its long, narrow neck and head stretched fourteen feet (4 m) and was especially shaped for reaching deep inside the large carcasses of dead dinosaurs.

Science tells us the pterosaurs all died out approximately seventy million years ago, during the great wave of extinction that swept the world at the close of the Mesozoic era. Yet the

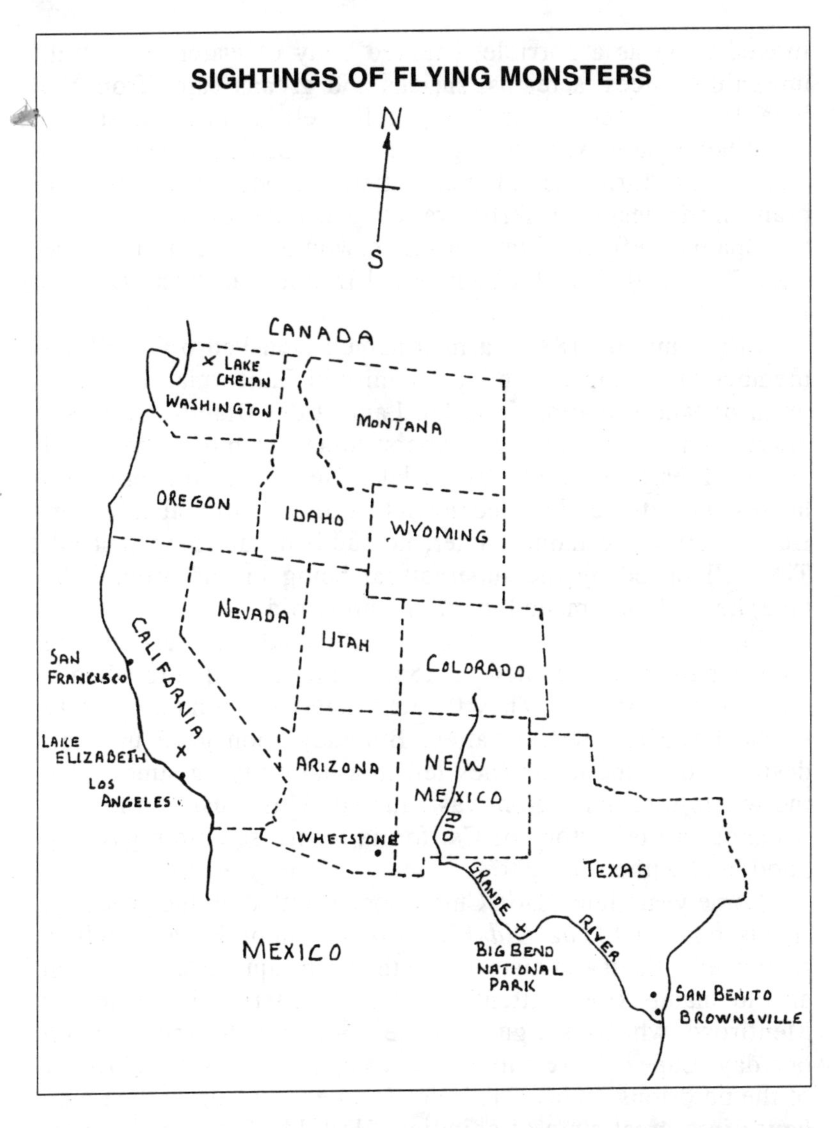

occasional report of a flying monster sighting in the United States, together with the odd disappearances of cattle and livestock, make us wonder. Could a few species have survived from the ancient past?

Reports of flying monsters go back to the time of the Spaniards in California. Lake Elizabeth had for centuries been looked upon as a horrible, haunted body of water. Frightful, unearthly voices, screams, shrieks and groans came from the lake. It almost seemed as though hell itself lay in its depths.

When Spain was issuing grants to Spanish citizens who came to California as colonial settlers, no one would accept a grant in the beautiful, fertile valley near Lake Elizabeth. After the Spanish left, the Mexicans also would not settle near the lake. They called it "La Laguna del Diablo," or "The Devil's Lagoon."

In the middle 1830s, a man named Don Pedro Carrillo, a member of the most prominent family in California, took up a grant of land radiating from the Laguna del Diablo. He was a brave man, not in the least superstitious. He built a beautiful hacienda, barn and corral by the lake. He stocked the land with horses and cattle and settled down to a life of baronial splendor. But, barely three months later, he suddenly abandoned it all. The hell raised by the supernatural being in and around the lake, he said, had made him prematurely old!

For over twenty years the beautiful lake and surrounding land remained unoccupied. In 1855, American squatters began coming to California. They thought they had found a paradise in the beautiful, fertile valley. But they soon gave up their desire to own the land. They left in a body saying simply that the whole infernal region was haunted. This must be the only instance in the history of California where squatters gave up good land without a fight!

Some years later, Don Chico Lopez settled on the property. In his book, *On The Old West Coast*, Major Horace Bell, a renowned citizen and historian of the times, quotes a story from an old manuscript written by a Don Guillermo Embustero y Mentiroso, who was a guest at the Lopez ranch. Around noon one day, Lopez's foreman, Chico Vasquez, said to be a brother of the notorious outlaw Tiburcio Vasquez, rode up to the ranch house in a great state of agitation. He told of strange happenings at the lake. Everyone immediately saddled up and dashed to the shore. They arrived to find the lake mirror-calm and serene. Don Chico Lopez immediately began berating the foreman for wasting their time on a wild goose chase. Suddenly there was a terrifying whistling, hissing scream from a thick growth of cattails at the edge of the lake. So close was the creature that they could smell its nauseating, foul breath. The men reeled in their saddles and the horses took off in fright.

As they brought the horses under control, the men turned and looked back. Silhouetted against the lake was a huge monster, larger than a whale. It had enormous bat-like wings which it flapped periodically as it tried to rise from the mud. As it roared, its huge flippers or legs churned the water to a foam.

The horses and men fled. Next morning all the vaqueros were mustered, armed and ordered to the scene. They found the lake once more serene. There was no sign of the monster. Had it flown away or sunk beneath the mud? they asked themselves. But the foul, nauseating odor still hung in the air.

In 1883, the Lopez horses and cattle began to disappear. At first grizzly bears were thought to be responsible, but the amount of devastation among the herds did not follow the pattern of grizzlies. One night there was a terrible uproar in one of the corrals. When the vaqueros reached the spot, ten mares and foals were missing. Outlined against the moonlit sky was the terrifying winged monster as it flew away, heavy with feasting. Don Chico Lopez promptly sold out cheaply to El Basquo Grande and left the cursed spot.

In October 1886, a Los Angeles newspaper carried a further report of strange happenings at the lake. According to the newspaper, a "python" or similar monster, terrible to behold, had made an appearance at Lake Elizabeth. The monster caused much terror and excitement among the local inhabitants. It reportedly fed on cattle, horses, sheep, pigs and chickens. At night the monster left the water and took its fill of sheep and calves—six at a time, from corrals.

On one occasion the monster tried to devour a Texas longhorn steer. The steer bellowed and kicked, attracting the attention of Don Felipe Rivera, its owner. He arrived on the scene to find the monster with the steer half sticking out of its mouth. The steer put up a terrific fight and eventually succeeded in freeing itself. Cheated of its meal, the monster made for the lake.

Don Felipe described the creature as about forty-five feet (13.7 m) long and as large as four elephants. Its head resembled that of a bulldog. It had six legs and wings which laid flat on its back when not expanded. Being a Don of true castilian blood, who feared not "hippogriffes, dragons, nor devils," Don Felipe pursued the python. As it floundered towards the lake he emptied his old fashioned forty-four caliber Colt into its side. The bullets, striking the monster's side, sounded as if they were striking against "a great iron kettle." One bullet bounced back and hit Don Felipe. Next morning he found four of the slugs flattened like coins by the impact.

Being an enterprising man, Don Felipe made plans to capture the monster and sell it to a circus. He signed a contract with Sells Brothers, who were to pay him $20,000 when he delivered the monster to them alive. Don Felipe's plans did not materialize, although the flying amphibious monster was seen several times between 1881 and 1886. El Basquo Grande, who bought the Lopez property, did pursue the monster on one occasion, but it entered the lake and sank without a trace. Later reports told of it emerging from the lake and flying eastward.

The monster never returned to the valley, but toward the end of 1886 a similar monster was found and killed 800 miles (1,287 km) away from Lake Elizabeth, in the Arizona desert. *The Epitaph*, a Tombstone, Arizona newspaper carried the story:

> A winged monster resembling a huge alligator with an extremely elongated tail and an immense pair of wings was found on the desert between Whetstone and Hauchuca Mountains last Sunday by two ranchers, as they returned home from the Huachucas. The creature was evidently greatly exhausted by a long flight and when discovered was able to fly but a short distance at a time. After the first shock of wild amazement had passed, the two men, who were on horseback and armed with Winchester rifles, regained sufficient courage to pursue the monster and after an exciting chase of several miles, succeeded in getting near enough to open fire and wound it. The creature then turned on the men but owing to its exhausted condition they were able to keep out of its way and after a few well directed shots the monster rolled partly over and remained motionless.
>
> The men cautiously approached, their horses snorting with terror, and found that the creature was dead. They proceeded to make an examination and found that it measured ninety-two feet (28 m) in length and the greatest diameter was about

> fifty inches (127 cm). It had only two feet, situated a short distance in front of where the wings were joined to the body. The back as near as they could judge, was about eight feet (2.4 m) long. The jaws were thickly set with strong, sharp teeth. The eyes were as large as dinner plates and protruded from the head. Some difficulty was encountered in measuring the wings as they were partly folded under the body. But finally one was straightened out sufficiently to get a measurement of seventy-five feet (23 m), making the total length from tip to tip about 160 feet (49 m).
>
> The wings are composed of a thick and nearly transparent membrane and are devoid of feathers or hair, as is the entire body. The skin of the body was comparatively smooth and easily penetrated by a bullet. The men cut off a small portion of the tip of one wing and took it home with them. Last night one of them arrived in this city for supplies and to make preparations to skin the creature. The hide will be sent to eminent scientists for examination. The finders returned to the kill early this morning, accompanied by several prominent men who will endeavor to bring the strange creature to town before it is mutilated.

Whether or not scientists were able to examine the hide of the pterodactyl-type monster has not been determined. Difficulties of transporting such a carcass or remains any distance without the aid of modern refrigeration, would be considerable. But, from such a detailed description, there can be no doubt an unknown flying creature of some kind was found by the ranchers in the Arizona desert. The likelihood of this monster being the same one originally seen on the Lopez property bordering Lake Elizabeth, is a possibility. But Don Felipe's description of a flying monster with six legs and a head like a bulldog does not quite fit the picture. There is always the possibility that Don Felipe, being an enterprising man with plans to capture the monster and sell it to a circus, might have embellished his story a little, particularly with a $20,000 contract with Sells Brothers riding on his description. Perhaps it was a different kind of flying creature entirely.

After the excitement of the Lake Elizabeth and Arizona desert sightings, nothing more was heard of flying monsters until five years later. On November 19, 1892, the *Bitteroot Times* carried another fascinating story concerning a monster strangely reminiscent of the prehistoric flying pterodactyl. The site of the incident was Lake Chelan, in north central Washington, on the eastern side of the northern Cascade Mountains, amid glaciated country.

Sixty-five miles (104 km) long, in a northeast-southeast direction, the lake is one to two miles (3 km) wide, has a surface elevation of 1,079 feet (329 m) and a maximum depth of 1,419 feet (432 m). The bottom of the freshwater lake is 340 feet (103 m) below sea level. The only shallows are near the outlet—the Chelan River, southeastward into the Columbia River. The Stehekin River is the principal feeder stream which heads in Cascade Pass and flows into the lake from the northwest.

The story concerned three travelers who were journeying to a point near the upper end of the lake. Upon arriving at a spot called Devil's Slide, one of the men approached the water and began his morning ablutions. As he was dipping his feet into the water for the first time, he suddenly cried out for help. Some unseen creature beneath the surface had attached itself to his foot and was dragging him into the water.

His companions rushed to his aid and taking him by the hands and arms finally succeeded in dragging him ashore. The *Times* goes on to say:

> ...but what was their surprise to see the monster also emerge from the water firmly attached to the man's leg by its teeth. It was a horrid looking creature, with the legs and body of an alligator and the head and restless eye of a serpent. Between its fore and hind legs, on either side, were large, ribbed feathery looking wings. The tail was scaled but not barbed like that in the picture of the typical dragon. With the exception of the under part of the throat and the tips of the wings, feet and tail, the creature was a beautiful white and its skin as soft as velvet. Knives, sticks and stones and everything else which were brought to bear upon the monster proved unavailing, and at last the ingenious travelers bethought themselves of heroic measure. They built a good fire and pulled the neck and belly of the beast, bird, or fish, across it, taking good care not to

> burn the leg of their comrade in the operation. After a while the scorching heat aroused the animal from its torpor. It began to move its body and to stretch out its leathery wings after the manner of a bat, and suddenly flew into the air, still holding the man by the leg.
>
> After rising to a height of about 200 feet (61 m) it took a "header" downward toward the lake, into which it plunged with a splash, burying itself and victim out of sight.

The newspaper concluded by saying the natives were greatly excited, believing that the great white dragon had reappeared and that the end of the world was at hand.

This is a remarkable story to say the least! It is also food for thought, in light of more recent reports. South Texas has repeatedly been the scene of "big bird" reports for decades. San Benito police had several sightings in 1975, 1976 and another in 1982. The most interesting of these occurred in February, 1976. While driving on a country road, a Texas rural school teacher was severely frightened by a gigantic flying creature that dived and swooped over her car. Its shadow covered the width of the road. To her added terror, a second flying monster appeared; possibly the mate of the first.

After the incident, the teacher searched encyclopedias at the local library for the creature's likeness. She found it among the prehistoric monsters. The creatures she had seen resembled a flying reptile believed to have been extinct for millions of years!

On the advice of her school principal she kept the matter quiet, since the news might create a panic among students and parents alike. But the matter was revealed when two other teachers also reported similar experiences. As with the first teacher, winged giants swooped low over their cars.

> It was the biggest thing I've ever seen alive, (one of the teachers told a reporter). Its just unreal. I don't know how it could have survived all those millions of years and still be flying around...It all happened so fast...it was such a shock you think you are seeing things. It was just enormous and frightening.

Soon there was a long list of reported sightings of the mysterious winged giants who seemed to have appeared from nowhere over southern Texas. No one knew what they could be. They were all simply dubbed as "Big Bird." Each day brought new reports into the offices of the Texas Parks and Wildlife Department. "We were fascinated," one state wildlife official remarked.

It appeared, however, the teachers were not the first people to see Big Bird. Just before dawn that day, while Police Officer Arturo Padilla of San Benito, an experienced hunter and outdoorsman, was driving his car, in the beam of the headlights he suddenly saw a huge broadwinged bird. The streets were deserted except for the occasional police cruiser. A short while later another police officer, Homer Galvan, saw a black silhouette of a giant bird gliding through the air. "It never did flap its wings," he told reporters. He estimated its wingspan was at least twelve feet (3.6 m).

The next report came from Alverico Guajardo, who lived in a mobile home on the edge of Brownsville. At approximately 9:30 P.M., while Guajardo was eating his evening meal, he heard a thumping against the side of his trailer. Opening the door cautiously, he was astonished to see what looked like an enormous bird standing in the half-light in his yard. He eased out the door and sped to his car. He switched on the headlights which revealed the most frightening creature the man had ever seen. "I was scared," he told a reporter. "It's like a bird but it's not a bird. That animal is not from this world."

Guajardo was most impressed by the eyes and the bill, which extended three or four feet (1 m). As it stood, bathed in the light of the car's headlights, the creature emitted a loud, rumbling bird call, such as the man had never heard before. He was able to study the creature from a distance of some fifteen feet (4.5 m), for several minutes before it finally walked away into the shadows.

Reports from along the Rio Grande Valley continued. Pictures of the footprint twelve inches (30 cm) long, found in freshly plowed soil, were shown on television. Excitement among southern Texans was at an all time high. A $1,000 reward was offered by a radio station for Big Bird's capture. A wealthy oil man offered another $5,000.

Guns were removed from racks and through the nights that followed men stalked the surrounding countryside. They searched dark hollows and the Texas sky with lights, for even a glimpse of the monster bird. The Texas Parks and Wildlife Department quickly became concerned. If what the reports indicated was true, Big Bird was certainly a very rare creature and therefore was on the endangered species list. As such, it was under full government protection and bounty hunters trying to kill or harm the creature were breaking the law and would suffer the consequences. The state wildlife officials were also concerned about other big birds of Texas. Vultures, sandhill cranes, brown pelicans, rare whooping cranes, wild geese—all were in danger of being shot by mistake by overzealous bounty hunters. "All birds are protected by state or federal law," the officials announced. This put a considerable damper on events from the hunters' point of view. After a period of two months and the absence of further sightings, the search gradually fizzled out.

Big Bird's true identity was never discovered. Some suggested it was an escaped South American condor, a brown pelican, a great blue heron. Others thought it was a relative of the prehistoric pterosaur and had hitherto gone undiscovered. Still others thought it was a creature from outer-space!

Whatever it was, the incidents along the Rio Grande brought to mind a startling discovery in 1971, in Big Bend National Park. While exploring a sandstone outcropping, Douglas A. Lawson of the University of California, discovered the remains of giant wing bones. He returned the following spring accompanied by his professor. It was agreed that Lawson had discovered the fossilized bones of a pterosaur. These were the first such bones ever found in Jordan and in the State of Wyoming.

In 1972 and 1974, two more specimens were found in the same area. Sixty million years ago Texas had been the haunt of the largest of all flying reptiles ever known to exist—quetzalcoatlus! The bones did not come from ancient seas. Instead, they came from hilly country far inland, more than 250 miles (400 km) from the nearest shoreline, where the giant creature lived, flew and found its food!

Whether Lawson's fossil had any connection with the Big Bird that later dived and swooped above the cars of terrified teachers and other people, we shall never know. But along the Rio Grande, people have not forgotten its mysterious appearance. Old-timers also wonder, for if the flying phantom did survive from prehistoric days to the time of the Old West, it could have survived in some desert fastness until today. There have been no significant changes of climate from that time to the present. Also, it might not be a coincidence that Mexico's Sierra Madre Oriental, one of the least explored regions of North America, lies only 200 miles (322 km) west, as the crow flies, of the San Benito/Brownsville area where the recent sightings occurred.

The traditions of the Indians favor the survival of at least some of the prehistoric creatures. Perhaps among these was an ancient pterosaur, the mysterious Flying Phantom of the Old West!

5

Cold-Water Denizens

Lake And River Monsters New England And Middle Atlantic States: Vermont, New York, New Hampshire

There are probably many more aquatic monsters than other monsters in the folklore of every people. The oceans, lakes and rivers are more mysterious and unaccountable than the surface of dry land. Reports, particularly of freshwater monsters, go back to the dawn of written history.

Perhaps the best known lake monster today is Nessie, the monster in Loch Ness, Scotland. As early as 565 A.D., the presence of monsters in Loch Ness was recorded. But it was not until 1960, when aeronautical engineer Tim Dinsdale shot his famous film strip, that Nessie was officially recognized by science. The film was analyzed and assessed by the Joint Air Reconnaissance Intelligence Center of the Royal Air Force. The Center is one of the few organizations in the world expert in interpretation of aerial photographs. Their report removed all doubt as to the existence of large, as yet uncaught and unidentified animals in certain deep, cold-water lakes that ring the Northern Hemisphere.

What are these cold-water denizens? And how did they get into the lakes? It is believed that lake monsters hark back to prehistoric times, when they formerly lived in the seas. The lakes were once fiords until the land level rose, sometime between the end of the last Ice Age and approximately 5,000 years ago. Trapped in the lakes, the "sea" monsters survived because the food and conditions suited them.

According to the late zoologist Dr. Ivan Sanderson, in size and shape, lake monsters are unlike any known aquatic animal. They seem to fall into two categories: tropical and northern. The tropical variety is usually described as reptilian, with scales and nose horns, much resembling the extinct reptiles known as theropods. The northern variety is a large aquatic animal generally in body shape, like the ancient plesiosaur of the Mesozoic period; more than seventy million years ago. The plesiosaur was a slender reptile, forty to fifty feet (15 m) long with a small head, long neck and two pairs of paddle-shaped limbs. It is into this category that most freshwater monsters are placed.

Like their cousins, the dinosaurs and pterosaurs, plesiosaurs were believed to have become extinct millions of years ago. Yet persistent reports of sightings of these creatures in oceans and lakes in Scotland, Sweden, Norway, Iceland, Northern Ireland, Scandinavia, Russia, Siberia, Canada and the United States, have forced scientists to revise their opinions.

North America's first residents, the Indians, had always known about the monstrous inhabitants of the lakes and held them in considerable respect. They spoke of them repeatedly in

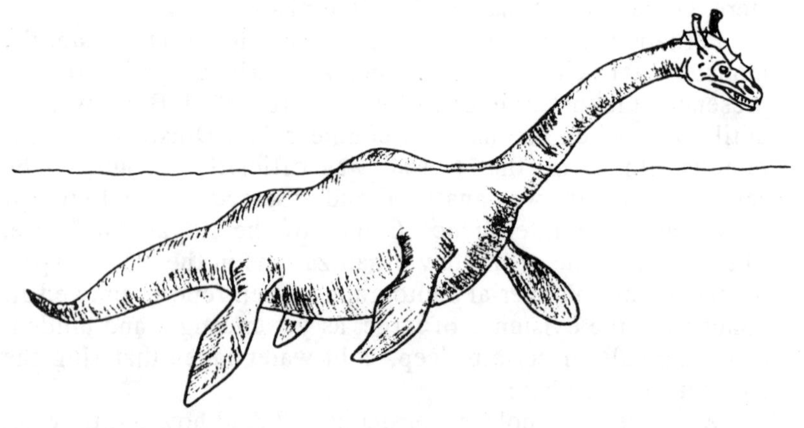

An artist's impression of the Loch Ness Monster.

their myths and legends. But nothing was documented until after Europeans began to settle the continent. Then, they too began telling stories of sightings and encounters with mysterious unknown creatures in the lakes and rivers. Over the past 200 years there have been literally thousands of documented sightings and much evidence has been accumulated.

Lake Champlain

A prime example of the long neck, or plesiosaur-type monster resides in New England. Here the landscape consists of low coastal plains and two rocky interior uplands; several mountain ranges and fertile river valleys separate the boulder-strewn uplands. In a great cleft, north to south in the terrain, separating Vermont from New York, lies Lake Champlain—the home of a venerable monster, aptly named Champ. Reports span three centuries.

Samuel de Champlain recorded his impression of the monster in July, 1609. He described it as a serpent-like creature "about twenty feet (6 m) long, as thick through as a barrel and with a head shaped like a horse's."

On July 24, 1819, pioneers in the Port Henry area were astonished to see the monster in Bulwagga Bay. In the 1870s it was seen near the Charlotte shore by a group of people on a steamboat. There were further reports in 1871, 1877, 1883/86/87/88 and 1894. Reports persisted into the new century with a flurry of sightings in the 1930s and 1940s.

In 1939, a couple quietly fishing in a small boat near Rousse's Point were forced to make a bee-line for shore when the monster suddenly appeared and began looping its way towards their boat.

The monster is usually described as being twenty to forty-five feet (6 to 14 m) long and is gray or rust colored. Its horse-shaped head sprouts two small horns. The neck is about one foot (30 cm) thick and four or five feet (approximately 1.5 m) long, with a thick, rough mane.

In 1945, Mr. and Mrs. Charles Langlois of Rutland, Vermont came close enough to the monster in a rowboat to "whack it with an oar." Mrs. Langlois was ill for several days after the frightening experience.

Various suggestions have been put forward as to the crea-

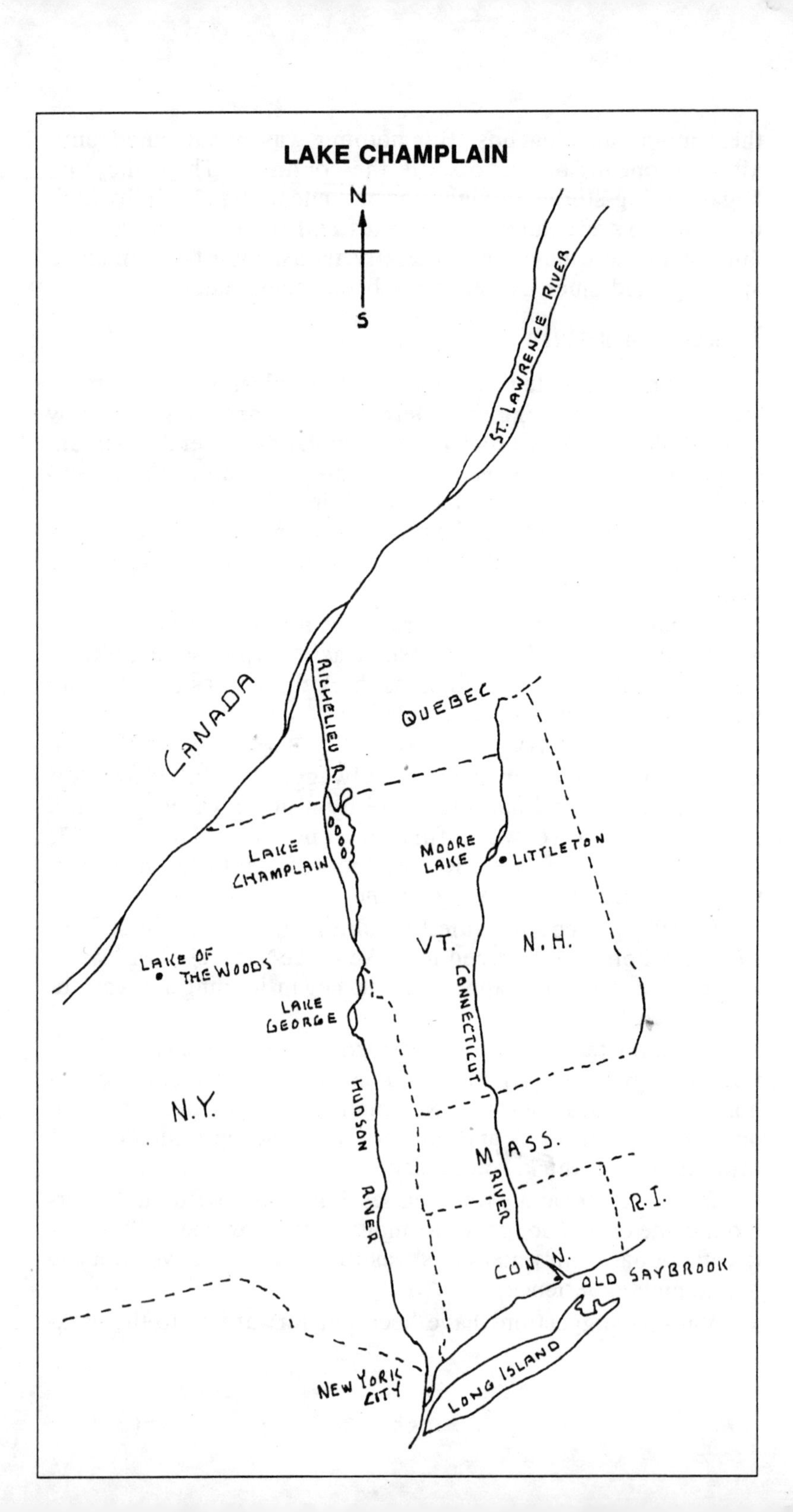
LAKE CHAMPLAIN
N
S
ST. LAWRENCE RIVER
CANADA
RICHELIEU R.
QUEBEC
LAKE CHAMPLAIN
MOORE LAKE
LITTLETON
LAKE OF THE WOODS
VT.
N.H.
LAKE GEORGE
CONNECTICUT
N.Y.
HUDSON RIVER
MASS.
RIVER
R.I.
CONN.
OLD SAYBROOK
LONG ISLAND
NEW YORK CITY

ture's identity. One scientist has suggested that the Lake Champlain monster may be some kind of giant slug which retreated to the depths of the lake centuries ago. Possibly human activity on and in the water frightens the creature and that is why it is seldom seen. Others have suggested the creature, or creatures, are really giant sturgeon "porpoising" in hot weather. But most favored is the theory that Lake Champlain's inhabitant is a plesiosaur-type animal or the Loch Ness monster's American cousin. Champlain's description and hundreds of others, bear a startling resemblance to Nessie and it may not be a coincidence.

There is an intriguing geographic similarity between Loch Ness and Lake Champlain. Loch Ness is the main deep-water lake in the Great Glen—a chain of rivers and lakes which knifes through the highlands and almost divides Scotland in two. Lake Champlain, together with Lake George and the Hudson River, forms a gap which likewise divides New England from the Adirondack and Appalachian highlands to the west. Climate and forests of the Lake Champlain area and the Scottish highlands are also similar. Lake Champlain, like the Hudson and St. Lawrence Rivers, was once an arm of the sea. Small prehistoric life forms are often found in their waters. New England takes her monster very seriously. The search for proof of his existence is continuous. As added encouragement to photographers to get busy with their cameras, the *Burlington Free Press* printed the following notice in its May 6, 1971 edition:

> Reward For Photo
>
> A bona fide and documented picture of the Lake Champlain sea-monster will mean $100—in color, $200—to some lucky lakefaring photographer.
>
> The *Burlington Free Press* offers that for a picture of the fabled creature seen by so many, yet never photographed.
>
> Believers date back 300 years, B.C. Before Camera. With today's equipment the elusive creature from the depths should not escape having his picture taken.

In 1977, a photograph of Champ was taken by Sandra and Anthony Mansi, which clearly shows an unusual creature with a small head, elongated neck and round body. The photo has been examined by experts, but unfortunately their findings were not positive—the photo is still largely considered suspect.

On April 18, 1983, the 150 member New York Assembly

unanimously approved a resolution "recognizing" Champ's existence and calling for his protection. The Assembly resolution is identical to the one passed by the New York Senate in June, 1982 and the Vermont House of Representatives, in April, 1982.

According to Joseph Zarzynski, researcher and founder of the Lake Champlain Phenomena Investigation, there have been over 300 sightings of this mysterious monster since the early 1600s.

Joseph Zarzynski at Lake Champlain during one of his Champ searches.
Credit: Pat Meaney

Lake-Of-The-Woods, New York

A similar Champlain or plesiosaur-type monster is purported to live in Lake-of-the-Woods, one of the numerous small and beautiful lakes in New York. It lies about seven miles (11 km) south of Alexandria Bay, on the St. Lawrence River, near the small town of Redwood. One morning in 1929, a lady and her mother were out fishing on the lake when the daughter noticed her mother's line jerking. She waited for her mother to reel in...

> When she didn't, I glanced up...she was staring awesomely at something in back of me. When I looked, I saw the most fantastic creature. We wasted no time pulling anchor and heading for shore—about 100 feet (30.5 m). It was quite close

> to our boat and both of us had the same fear—it might upset us. Neither of us spoke until we climbed out on the dock. Then a reaction set in. We still didn't believe what we had seen.
>
> This water creature appeared to be about twenty feet (6 m) in length...grayish tan, with a head not much bigger around than the neck or front loop of the body; but with a saw-tooth growth (possibly a mane) along its head and down the length of the neck or body, about six or seven feet (approximately 2 m). It was looking our way and about three arcs were plainly visible. When we started rowing it submerged, causing the calm surface of the water to become agitated.

The report came as no surprise to two brothers, owners of half the area around Lake-of-the-Woods. They had seen the creature a number of times and had once found a sort of slide on the bank, possibly where it had come ashore. Further enquiries revealed that other visitors to the lake had also seen the creature.

Moore Lake, New Hampshire

A strange and eerie monster story came from Moore Lake, in 1968. About 3:00 A.M., May 20, three "badly frightened" young people, Mr. and Mrs. Richard Hansen and nineteen-year-old Michael Stinchfield, burst into the Littleton Police Station shouting about a "red glow on the water," and a "thing" that appeared while they were fishing. Apparently, earlier in the evening they had driven out to Samuel C. Moore Lake to fish for brown bullhead. There was no moon and the lake lay dark and quiet. Fishing was slow, but they didn't mind as the night was pleasant.

Shortly after 2:00 A.M., Stinchfield pointed to a red glow on the water about a quarter of a mile (400 m) north of where they stood. The glow was partly obscured by a rock ledge extending out into the lake. They thought it strange but soon forgot about it.

Soon, they became aware of an eerie silence that had crept over the night—no frogs croaked, no animals moved in the woods. They continued casting lures into the water and reeling them back listening to the gurgling sound the metal plugs made in the strange stillness.

Suddenly Hansen cried, "Look at that!" The red glow had moved from behind the rock ledge and was now about thirty feet (9 m) out from the wharf where they stood. It seemed to come from something lying motionless in the water. The thing appeared to be a "whitish mound" about two feet (61 cm) wide, extending about a foot (30 cm) above the surface. Just above the waterline were two round markings, which looked like red glowing eyes. Something larger loomed from the darkness behind the mound.

Stinchfield said the object looked like an alligator submerged up to its eyes in the water. But the Hansens felt the red glow and the darkness made it impossible to give a likeness. Mrs. Hansen and Stinchfield were quick to leave the wharf, but Hansen stayed. On an impulse, he cast his lure out towards the thing as it lay, motionless and quiet, in the water. As he reeled in, the thing suddenly rushed towards the wharf, making a sound like the bubbling of an aqualung under water.

Mrs. Hansen screamed. Her husband dropped his fishing rod and the three raced to their car. They did not stop to look back but started the engine and headed down the road. Just before turning a bend in the road, Hansen stopped the car. As they looked back the area around the wharf seemed to be glowing red. The young men's fear now changed to curiosity. They suggested driving back for another look. Mrs. Hansen demanded they return to town immediately. "For a week after that I couldn't look at a red traffic light or neon sign at night without beginning to shake," she said.

At 3:00 A.M., accompanied by Officer Miller, they returned to the wharf. The red glow was gone but Hansen's fishing rod still lay as he had dropped it. All four noticed the eerie silence that lingered about the lake.

A reporter who later interviewed the young people, suggested the thing might have been a deer swimming in the water. But all three quickly disagreed. Suggestions that the thing might have been a flock of loons, a large turtle or a pike, were also rejected by all three witnesses. While being skeptical of monsters, Chief of Police Stanley L. McIntyre was nevertheless puzzled. Both Stinchfield and Hansen were experienced woodsmen and didn't "seem the type to scare easily."

Other odd things which added credibility to the young people's report were also reported from Moore Lake. Chief McIntyre's investigations in daylight revealed horned pout (the local name for brown bullhead), strewn along the shore near the wharf and strangely, only the heads, tails and spines of the fish remained.

Since the Hansen's and Stinchfield's report, other people, including a shop teacher in Littleton's Public School, reported seeing red lights on the lake that night. Another nearby resident, Roger Caswell, also reported noting the eerie stillness over the lake on the night of May 19–20. There were other rumors of strange glowing red lights in the area, but no one else spotted the thing which so unnerved the young people.

Whatever the creature was, it could have come up from the ocean. The Connecticut River, which begins in Quebec, flows down between Vermont and New Hampshire, snakes through Massachusetts and Connecticut and empties into the Atlantic Ocean at Old Saybrook. In the early 1960s Moore Dam was built at Littleton, forming a very deep lake some eleven miles (18 km) long and one mile (1.6 km) wide. Some creature could have made its way up from the ocean before the dam was built and subsequently became trapped when the lake was closed.

Perhaps the creature's identity will never be learned, but an aura of mystery still surrounds Moore Lake, especially on warm summer nights when an unexpected eerie stillness creeps over the lake.

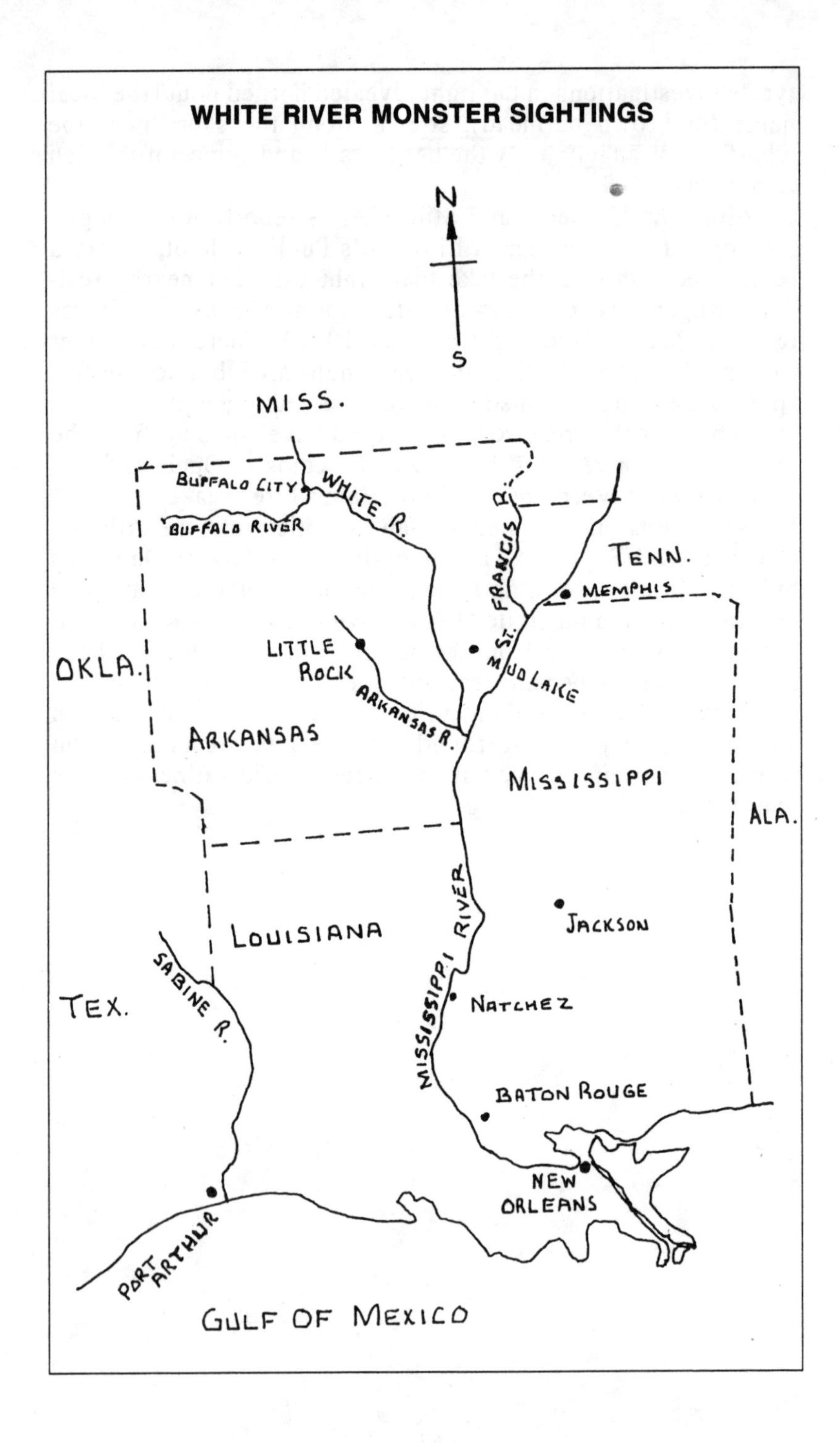
WHITE RIVER MONSTER SIGHTINGS
N
S
MISS.
BUFFALO CITY
WHITE R.
BUFFALO RIVER
ST. FRANCIS R.
TENN.
MEMPHIS
LITTLE ROCK
MUD LAKE
OKLA.
ARKANSAS R.
ARKANSAS
MISSISSIPPI
ALA.
MISSISSIPPI RIVER
JACKSON
LOUISIANA
SABINE R.
TEX.
NATCHEZ
BATON ROUGE
NEW ORLEANS
PORT ARTHUR
GULF OF MEXICO

6

"Whitey," The White River Monster

Lake and River Monsters Southern States:

Arkansas; Florida

An interesting assortment of monster reports spanning more than 150 years have come from the southern states. It is beautiful country of gently rolling hills; sandy beaches stretch along the coastal plain and large forests of pines and other trees spread over about half the region. Most of the early reports were from southeastern Arkansas.

Mud Lake

It is not often that a monster is captured, but *The Forrest City Times*, May 28, 1897, reported such an event, at Mud Lake in the St. Francis bottom. Apparently, each evening, residents on the borders of the lake were startled by an unearthly coarse whistling note mingled with loud slapping of the water and a great commotion. In the middle of all this, a glimpse now and then revealed the outlines of an aquatic monster, such as was never seen in those parts. The performance was regularly repeated about the same time in the afternoon and always attracted a full house from local residents around the big lake. The *Times* went on to say:

> A council of the hardy natives was held and various means were suggested leading (to) the capture of what they termed the sea serpent...it was agreed, upon the urgent recommendation of an old salt who had seen service among whalers way down in the Indian Ocean, to try the harpoon process. Two of these instruments were turned out of a plantation blacksmith shop, which equipped the attacking party for a hand to hand encounter. At the appearance of the monster, two boats were speedily manned and bore down upon the aquatic flounderer while he was seemingly enjoying his afternoon bath.
>
> With great caution, the two boats neared the object of their murderous intention. So engrossed in his mode and mood of hilarity (churning the water and splashing it high in the air), the monster was unconscious of approaching danger till the deadly 'poons sunk deep into vital parts of his body. Simultaneous with the strokes, which fastened the instruments into the flesh of the strange visitor, a desperate lunge was made for liberty and deeper water, which sped the boats over the surface of the glassy water at a terrific rate.
>
> The crews were determined and plucky and played out line and took up slack, as their swift and perilous cruise demanded.

Fortunately, no part of the lake is very deep, which was decidedly in favor of the attacking party. Some two hours later, the monster was finally hauled out on the bank by teams. It was sixteen feet (5 m) long and eight feet (2.5 m) around the thickest part of the body and weighed as much as "the largest ox." The body was covered entirely by a thick, scaly skin. In the massive jaws were three tiers of dagger-like teeth that curved inward. Nothing like it had ever been seen in the area before. The monster's appearance at the time was attributed to a recent flood.

The astonished local residents preserved pieces of the anatomy and there was enough food in the fleshy parts of the monster to feed the swine for several days. *The Forrest City Times* also added:

> His fish majesty had evidently lingered through a portion of

the iron age as his exterior was decorated with two gigs and a hatchet.

White River, Arkansas

Reports of a monster in the White River have occurred in forty year intervals since 1850. This sizeable river flows through eastern Arkansas and empties into the Mississippi River, which in turn runs into the Gulf of Mexico. At its deepest point, near Jacksonsport, the White River is over 100 feet (30 m) deep.

According to *Pursuit*, Vol. 4, 1971 and an expert on the folk history of north central Arkansas, the White River monster traveled "clear up into the Buffalo River." During World War II, "what some folks feared were German submarines coming up the Mississippi River as far as Memphis (well north of the mouth of the White River), were really sightings of the dreaded White River monster coming home after a jaunt to the ocean."

Little attention was paid to early reports of a huge creature lurking in the "deep hole," until 1937, when prominent citizens began reporting sightings. The creature was generally described as "big as a boxcar," and "as wide as three automobiles" rising from a sixty foot (18 m) eddy in White River, just a few miles below Newport.

The reports generated tremendous excitement. Hardly a city or hamlet across the country had not heard of Newport's monster. The Chamber of Commerce decided to investigate. A diver named Charles A. Brown, who worked for the United States Corps of Engineers, was hired and sent down into the murky water to search out the monster's hiding place. Much publicity was given the operation. Reporters and movie camera men descended on the scene. The Chamber of Commerce set up a dance hall on the river bank and booths selling cold drinks and sandwiches for the hundreds of people who flocked there to watch. On one day, scores of cars from as many as seven different states lined the river bank. It was all great publicity for the town. After several days of fruitless diving operations, hampered by poor visibility and the extremely muddy condition of the water, the search was abandoned. Brown did establish, however, that contrary to the theory of local skeptics, the monster was not a sunken boat, gas formation, logs or sunken drift.

He had searched the bottom of the river where the monster was seen and found nothing. In his opinion, some large creature did live in the depths, but had now moved on, up or down the river which was at its high stage.

Besides the large number of reports from local residents, letters poured in to the Chamber of Commerce from other sources. All stated they had seen something they could neither explain nor understand. Some reputable citizens such as J. M. Gawf, a local merchant, planter and stock raiser, even signed affidavits:

> ...on or about the first of July, 1937, Jess Outlaw and myself went to the scene of the supposed "monster" and while standing on the bank of the river saw a disturbance in the water across the river some 300 feet (91 m) from us. The water would come up something like two feet (61 cm) and would boil out. The waves would reach 100 yards (91 m) up and down stream.

While Mr. Gawf and Mr. Outlaw did not see the creature, they were certain no known animal could have caused such turbulence.

Another long-time resident and wealthy landowner also signed an affidavit:

> I, Bramblett Bateman state under oath, that on or about the first of July, 1937, I was standing on the bank of White River about one o'clock and something appeared in the river about 375 feet (114 m) from where I was standing, somewhere near the east bank of said river. I saw something appear on the surface of the water. From the best I could tell, from the distance, it would be about twelve feet (4 m) long and four or five feet (1 or 1.5 m) wide. I did not see either head nor tail but it slowly rose to the surface and stayed in this position some five minutes. It did not move up or down the river at this particular time, but afterward on different occasion I have seen it move up and down the river, but I never have, at any time been able to determine the full length or size of said monster.
>
> Some two weeks ago...September 22nd, 1937, I saw the same thing up-stream about 200 yards (183 m) from where it made its first appearance. On the last date that I saw the monster it

> was in the current of the river. Before it was always seen in the eddy. There is no question in my mind whatever but what this monster remains in this stretch of water as was first seen.

Mr. Bateman further stated that he could secure sworn affidavits from at least twenty-five other people who had seen the monster. Mrs. Bateman likewise signed an affidavit to having seen "the water boil up across the river, about two feet (61 cm) high."

Z. B. Reid, Jackson County Deputy Sheriff, signed another affidavit stating that shortly after 6:00 P.M. on the last Saturday of June, Deputy Joe McCartney, Henry Harper and himself saw:

> ...a lot of foam and bubbles coming up in a circle about thirty feet (9 m) in diameter some 300 feet (91 m) from where we were standing. It did not come up there but appeared about 300 feet (91 m) upstream. It looked like a large sturgeon or cat-fish. It went down in about two minutes.

After the initial excitement, things quietened down and nothing was heard of the White River Monster for thirty-four years.

In mid-June, 1971 an incredible story was told to a *Newport Independent* reporter by a prominent citizen who wished to remain anonymous:

> I don't want you to think I'm crazy, but I just saw a creature the size of a boxcar thrashing in the White River...It didn't really have scales, but from where I was standing on the shore, about 150 feet (46 m) away, it looked as if the thing was peeling all over. But it was a smooth type of skin or flesh.

He went on to describe the monster as about the length of three or four pickup trucks, and at least two yards (2 m) across. Before it appeared, the water began to boil up about two or three feet (1 m) high.

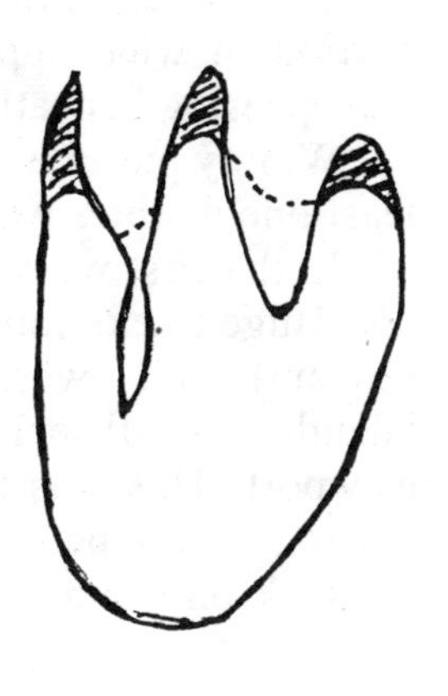
"Whitey"—White River, Arkansas, 1971

> ...then this huge form rolled up and

over; it just kept coming and coming until I thought it would never end.

Shortly thereafter, "Whitey," as the people of Newport had now named the monster, was in the news again. One witness, Earnest Denks, described him as a huge, gray creature probably weighing over 1,000 pounds (454 kg) with a long pointed object protruding from his forehead and looking as though he came from the ocean.

On June 28, three Newport fishermen saw him south of the White River bridge and about 200 feet (61 m) away from their boat. One of the fishermen, Cloyce Warren, took the only known photograph of the monster. He gave the following report:

> I didn't know what was happening. This giant form rose to the surface and began moving in the middle of the river, away from the boat. It was very long and gray colored. It appeared to have a spiney backbone that stretched for thirty or more feet (9 m). It was hard to make out exactly what the front portion looked like, but it was awful large. It made no noise except for the violent splashing and large number of bubbles that surrounded it...The creature looked like something prehistoric. The tail was constantly thrashing, and bubbles and foam surrounded the upper part, or I should say the front.

The picture came out but, unfortunately, it shows only a portion of what appears to be a gray-skinned moving object disappearing beneath the surface.

Whitey put in a number of appearances that summer and at least one doubter was converted overnight.

In the first week of July, there was another strange discovery. Huge tracks fourteen inches (36 cm) long and eight inches (20 cm) wide, were found by three fishermen on Towhead Island—a small secluded area about six miles (9.7 km) south of Newport. This was the same area where Whitey had been cavorting for the past three weeks!

Jackson County Sheriff Ralph Henderson, was soon on the scene with nine other men. Game Warden Claude Foushee shook his head and declared: "I've never seen anything like these huge tracks." State Trooper Ronnie Burke agreed. Besides their huge size, each track had three toes with claws, large pads on the heel and toes, with a spur extending at an angle from the heel of the print.

While the sheriff made plaster casts of the prints, a psychologist named Mike Loos explored the island. A sandy beach ran along one side of the island, the rest was dense foliage. At the far end of the island, approximately a quarter of a mile (.4 km) away, Mike Loos came upon another set of tracks leading from the water. Many small trees were pulled over and a large area of grass was flattened, as if something heavy had been lying there. He discovered where the tracks came from and returned to the water. The distance between the prints was eight feet (2.5 m)!

The weary and the thoroughly mystified investigators finally left the island. They could not rule out the possibility of hoax. Yet, if this were so, then the pranksters had gone to a lot of trouble and had put the tracks in an unlikely spot. The psychologist was most bothered by the distance between the prints and the fact of their being in such a secluded island.

Where the people of Newport were concerned, the tracks had to be the signature of their weirdly elusive, half-legendary monster, Whitey. As far as anyone knew, this was the first time he had left any positive evidence of his existence.

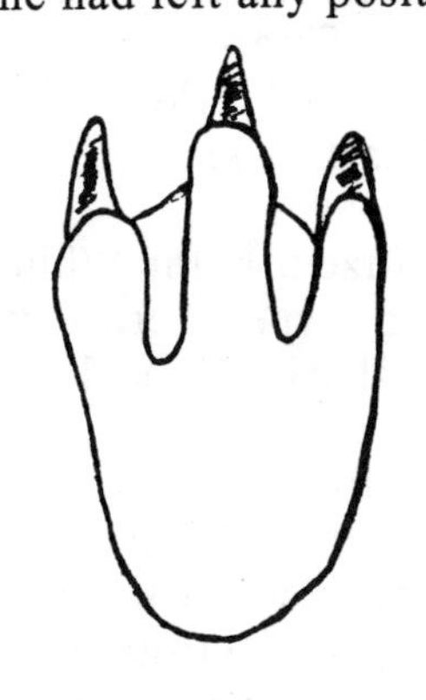

Penguin tracks

The sheriff figured the tracks were about three weeks old; significant with the monster's recent sightings and adding a new aspect—Whitey sometimes came out of the water! Many a staunch Newport citizen now cast an uneasy glance over his shoulder after sundown along the banks of the White River.

Experts finally concluded that Whitey was a truly gigantic penguin. Paleontologists from New Zealand had recently uncovered such a fossil in their country.

Witnesses had described the monster as "smooth and gray and very large," and as "rolling over and over," so that any spines would be noticeable. They also said it looked as if it were "peeling all over, but had a smooth type skin or flesh." Penguin feathers flake away like the sloughing of a reptile's skin. Several witnesses mentioned the long, pointed object protruding from the creature's forehead. This could in fact have been the penguin's beak.

Consideration was also given the fact that similar tracks to the White River ones had been discovered in Florida, in 1948, when a giant creature of some kind had apparently come out of the water at Clearwater and meandered along the western Florida shore and eventually up the Suwannee River. In its wake were huge, webbed three-toed footprints eighteen inches (46 cm) long and extraordinarily like those of a gigantic bird. From their depth in the damp sand, it was estimated this incredible biped must have weighed two or three tons and stood about fifteen feet (4.5 m) tall.

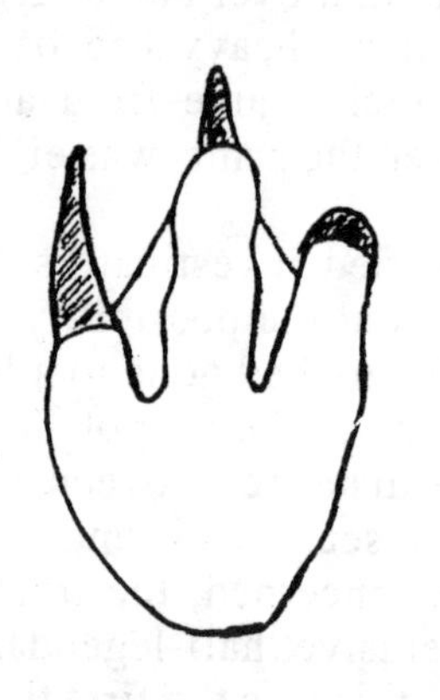

"Old Three-Toes"—Suwannee River, 1948

Here also, it was the opinion of the experts that the tracks were made by a giant bird of the penguin type, because of the disposition of the claws and their comparative lengths with those of the toes. At the time of the incident "Old Three-Toes," as the monster was called, created quite a stir in Florida, with sightings coming from as far as fifty miles (80 km) up the Suwannee River.

In light of this information, concerned citizens in the White River area began to wonder whether the same type of creature might also be loose in Arkansas. After all, if Old Three-Toes could get fifty miles (80 km) up the Suwannee River, it seemed perfectly possible that others could travel from the Gulf of Mexico far up the Mississippi and its biggest tributaries!

However, their fears were unfounded. Seventeen years later Tony Signorini, one of the partners at Auto Electronics in Clearwater, admitted the "giant penguin" tracks that mysteriously appeared on a Clearwater beach in 1948 were a hoax. The perpetrators were Signorini himself, and well-known pranskter Al Williams. Signorini kept the secret for forty years and only admitted to the hoax after his partner's death in 1988. The tracks were made by a pair of cast iron feet attached to high-top sneakers! The two men evidently thought it jolly clomping around on the beach hoodwinking the public. It seems many people including the police, believed the whole thing was a hoax, but they had no way of proving it until Signorini came forward. While the whole "Florida three toes" episode is best forgotten, Whitey still manages to keep public interest.

Lake Conway, Arkansas

While any monster would find it hard to compete against the White River monster, others did cause momentary flurries in the region from time to time. In 1953, the focus was on Lake Conway, near Forrest City. For more than three decades fishermen had been telling stories of strange finny monsters in the deep, dark recesses of the lake. One fisherman claimed he dragged a freakish animal of about eighty pounds (36 km) to the surface on a trot line. When he tried to pull it towards his boat to have a closer look, the creature suddenly surged away and disappeared beneath the dark surface.

A carpenter who lived east of the lake and about six miles (9.6 km) south of Saltillo, said while he was casting for bass in shallow water he noticed a stirring motion some yards away. Thinking it was fish feeding, he silently approached. The motion suddenly came towards his boat, passed it to a distance of about forty feet (12 m) and then returned. About twenty-five feet (7.6 m) away the disturbance suddenly ceased and a "head" came to the surface in some weeds. Then the body appeared. It was larger than a man, hairless and with a "dark brown frog-like skin." No ears or eyes were visible. The animal remained perfectly still as the man took aim and fired his thirty-two caliber pistol. The bullet went high, but the animal did not move. The man fired again. There was a momentary thrashing about in the weeds and the creature disappeared.

The fisherman watched and waited for a few moments and when it did not surface, he moved his boat to the place where it had gone down. Dipping his paddle to the bottom, the man stirred it about. There was a vigorous motion as his paddle came in contact with the creature. In a swirl of water it shot away. There was no sign of blood in the water, but thinking he might have wounded or killed the creature the man returned with a friend to the spot later in the day. They hoped to find the carcass and bring it back. There was no sign at all of the creature. The man said he had fished that spot before, but had never seen anything like the animal.

There was another incident on June 5, 1953, at Brannon's Landing. Joe Mosby, who wrote a column for the *Arkansas Gazette*, reported a more recent sighting that was told him by a relative of a *Gazette* employee. In this case, the creature "splashed around near the man's boat and raised up out of the water. It was a dark color, probably brown, marked by noticeable orange spots. Its neck was rounded and with little or no visible neck."

St. John's River, Florida

In June 1975, the *Arkansas Gazette*, carried a report of a monster seen by a fishing party in the St. John's River, Florida. "Pinky," as the monster was nicknamed, sounded more like a character from a Walt Disney movie. He was seen one Saturday by a fishing party—Mrs. Dorothy Abram and her husband Charles, owners of Harrell Glass and Supply Company at Jacksonville; Brenda Langley and two friends. About 10:00 A.M., due to an approaching storm, the party decided to head for shore. They started the motor, the boat turned and suddenly, to their astonishment, about twenty feet (6 m) away, there was Pinky!

> His head was about the size, or maybe larger, than a human head (Mrs. Abram recalled). He turned his head like you and I. He had...horns...like a snail with little knobs on the end. He looked like a skeleton. He was real jagged looking. On the side of his head were flaps, I guess some kind of gills, hanging down. About three feet of his neck was out of the water. His mouth was turned down at the edges and he had big, dark, slanted eyes. And he was pink. Sort of the color of boiled shrimp.

Mrs. Abram was not the first of the party to see the creature. Brenda Langley had seen him two or three times earlier in the day, but nobody had believed her.

In all, Pinky watched the people for about eight seconds, the witnesses said, before he quietly submerged and headed for the ocean. Their party headed for shore—fast!

The Florida Marine Patrol told Mrs. Abram that they had a lot of calls from people seeing things and always replied that they were probably sea sturgeon because they do grow up to eighteen feet (5.5 m) long. But Mrs. Abram did not agree. A sturgeon has a pointy nose. The creature she saw had a neck and a head. It was turning its head like a man would when looking around. She further stated, "it looked like a dinosaur with its skin pulled back so all its bones were showing."

Since their sighting, Mr. Abram said they had received several calls from other people who had seen the creature. One woman said she and others had seen it twenty years before, but no one believed them. A gentleman from Atlanta called and described almost exactly what the Abram party had seen. He had also seen the creature previously.

Jeff Hallett of Marine Science Center at Mayport, near Jacksonville, suggested three possible explanations for Pinky: a sturgeon, a sea cow or merely a floating tree trunk, giving the semblance of a monster. He admitted, however, that there are very few sturgeon in those waters and that a tree stump would not "move its head" as Pinky was purported to have done. Also, sea cows—big walrus-like mammals who weigh up to a ton, don't have long, skinny, bony necks. He had to admit he was "kind of fascinated" by the reports.

"Will Pinky now join the list of monsters we have known and loved?" asked the *Gazette*. "Or will he just sit there and sulk in the deep blue waters? If he's to be a full fledged major monster, someone will have to get the word to him: he's got to appear more often!"

K.W.

7

"Bozho," The Madison Monster

Lake And River Monsters Midwestern States: Indiana; Wisconsin; Mississippi; Minnesota; Michigan; Nebraska

The midwestern states cover the northern part of the great Interior Plain region and stretch from the Appalachian Mountains in the east to the Rocky Mountains in the west. They border Canada in the north and reach as far south as the northern border of Arkansas; an area equivalent to nearly one fifth of the area of the United States.

A large number of monster reports have come from this area, some dating back to the early nineteenth century. One of the earliest reports was from Indiana.

Lake Manitou, Indiana

The northern part of Indiana is dotted with hundreds of small, clear and beautiful lakes. Among them is Lake Manitou, seventy-five miles (121 km) from Logansport, near the town of Rochester. The lake is about two miles (3 km) long, half a mile (.8 km) wide and of unknown depth. Soundings were once tried with a line the surprising length of forty fathoms (240 feet), but with no effect. Known by the Pottawattomi Indians as "The Devil's Lake," Lake Manitou has been part of their legends for centuries. In its unfathomable crystal depths, they believed, lived the Evil Spirit. So strong was their fear of the lake in early times that they would not hunt, fish, nor bathe in its waters for fear of incurring the anger of the evil spirit.

The *Logansport Telegraph*, July 21, 1838 carried the first newspaper report of a monster in the lake, although sightings had been made several times over the previous ten years. The monster was seen by some men named Robinson while they were fishing near Rochester. They were surprised by something approximately sixty feet (18 m) long, swimming rapidly and creating quite a disturbance in the water. Not being easily frightened, they were nevertheless disturbed enough to immediately return to shore. A few days later a man named Lindsey, a well-respected member of the community, was riding near the lake. Some 200 feet (61 m) from the shore he suddenly noticed an animal's head raise three or four feet (1 m) out of the water. He watched the mysterious creature for several minutes, during which time it disappeared and reappeared three times in succession.

He described the head as about three feet (1 m) across the frontal bone and having a contour of a "beef's head." The neck was tapered and serpent-like. Its color was dingy with large bright yellow spots. The head turned with an easy motion, as though the monster were surveying its surroundings.

Stories of the monster flourished; so much so, that it became hard to separate fact from Indian legend. So many people believed the monster stories that an expedition was proposed to the lake. With the aid of rafts it was hoped the mysterious monster, "a terror to the superstitious," but "an object of interest to science, the naturalist and philosopher," would soon be captured.

> It is astonishing (said the *Telegraph*) that such a small inland lake, so remote from the seas, should be as mysterious in its depths as it is in its legendary associations. But so it is. Boys! Up with your harpoons and to the Lake Man-i-too!

Despite the rallying call, no expedition seems to have taken place; possibly "a sickly season combined with other circumstances" discouraged the volunteers from their purpose; or perhaps the growing tension at the time between the Pottawattomi and the white men made such an undertaking seem inadvisable.

Rock Lake, Wisconsin

Wisconsin lakes fairly teem with monsters! Reports span more than a century. One of the earliest newspaper reports was printed in the *Lake Mills Spike*, later re-named *The Leader*, on August 31, 1882. It concerned a "sea serpent" in Rock Lake.

> Again has the lake monster been seen. On Monday evening last, as Ed McKenzie and S. W. Seybert were rowing a race out near the first bar they discovered on the surface of the water, a little in advance of their boats, what they supposed for a moment to be a floating log...the latter called to the former to look out and not run into it...at this moment the object manifested life and reared its head about three feet (1 m) out of the water, opened its huge jaws about a foot (30 cm) or more, and dived out of sight. Almost immediately its head was thrust a couple of feet (61 cm) into the air close beside Ed's boat. "Strike him with the oar," yelled Mr. Seybert. But Ed screamed with terror, stood up in his boat and called ashore for help...
>
> Captain Wilson seized his shotgun and jumped into his swiftest boat and soon reached the frightened rowers. He saw the place at which the monster went down. The air all around was heavy with a most sickening odor. Ed was white as a sheet, his teeth clattering...John Lund said he could from shore, distinctly see the animal. He said at first he thought it was a man struggling in the water. Ol Hurd, from the same position, thought it was a huge dog.
>
> Ed McKenzie says it was fully as long as his boat, and somewhat the color of a pickerel. He says, "Let them talk about striking it with an oar, or anything, there isn't one of 'em would do it if they'd seen it come up sudden like, with its mouth wide open."

The monster has quite a history. In 1867, a Mr. R. Hassam saw it in the rushes. He first mistook it for a tree limb, but upon closer inspection saw it was a creature of some sort. He struck it with a spear but "could no more hold it than an ox." Mr.

Harbeck of Waterloo, who formerly resided across the lake, saw the saurian frequently while rowing back and forth. On one occasion it raised its head and hissed at him as he was entering the rushes. Another man named Fred Seaver, had two encounters with the creature; once it seized his trolling hook and pulled his boat over half a mile (800 m) at a rushing speed before it let go.

On another occasion it seized John Lund's troller. In his attempts to bring the creature in, the line cut deeply into his finger, before breaking.

Not one or two, but many people reported seeing the monster raise its head from the waters. "Perhaps it was a giant species of extinct amphibian origin," the *Lake Mills Leader* commented later.

MADISON'S FOUR LAKES

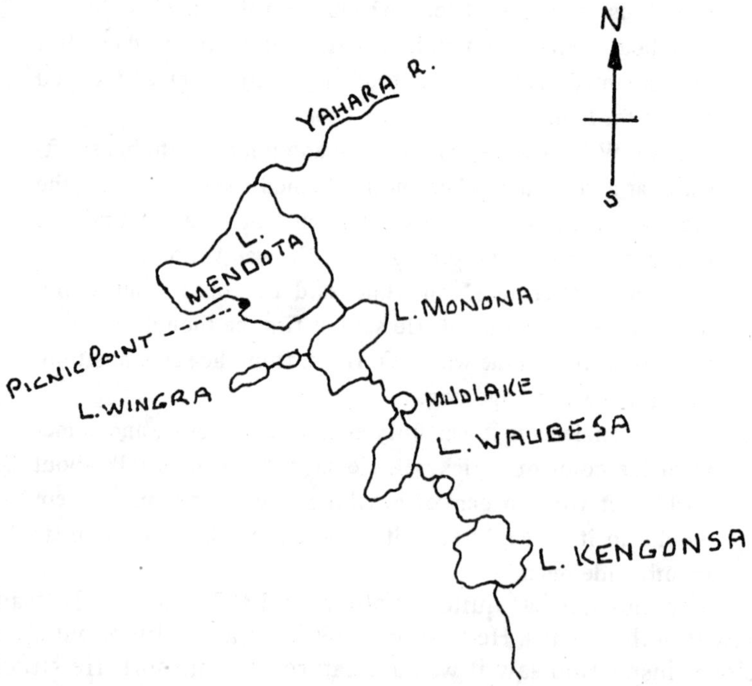

Lake Mendota, Wisconsin

One of the most famous of Wisconsin's monsters is "Bozho," the Madison monster, who lives in Lake Mendota, the largest of the chain of lakes known as The Madison Four Lakes. Prior to 1883, the monster was seen on numerous occasions and its existence was championed by many respectable citizens. But to Billy Dunn alone—one of Madison's most famous fishermen, was accorded the honor of hand-to-hand combat with the monster.

It happened on a warm day in June, 1884. According to the *Tribune*, July 24, 1892, Billy Dunn and his wife were quietly fishing near Livesey's Bluff, in Lake Mendota—the fairest of Madison's four lakes, when he noticed a black object moving threateningly towards the boat. As it came nearer, he saw it was a huge snake with reptilian head raised several feet above the water. Its forked tongue darted fiercely backwards and forwards. The surrounding water was considerably disturbed.

> Dunn was equal to the occasion and seizing an oar awaited the attack. The serpent, with a fierce hiss, sprang upon the boat, but only to fall back partially stunned by the well-aimed blow of the fisherman. The oar could not be withdrawn however, before the coils of the snake had surrounded it with their firm embrace...the reptile, recovering from the blow, had darted its long black fangs entirely through the blade. Now was Dunn's opportunity and as the monster was struggling to disentangle its teeth from the wood, he rained blow after blow upon it with a hatchet he carried by his side, until the snake gave up the contest, uncoiled itself and sank beneath the waters.

The *Tribune* went on to say that Dunn kept the oar, with some of the huge black fangs embedded in it, as a memento of his terrible experience. The man made no guess at the creature's length, but said its color was light greenish with large white spots. The *Tribune* omitted to say what Dunn's wife was doing throughout his terrible ordeal. Perhaps she had fainted from shock!

In 1917, some thirty years later, a University of Wisconsin student found on the beach at Picnic Point's north shore an object that resembled a fish scale. It was large, thick and very tough. Never having seen anything like it before, he showed it to his professor. He was from New England and was acquainted with the species. He identified the scale as that of a "sea serpent." This seems to be the first well verified indication of such a monster in Lake Mendota.

Nothing was seen of the monster until the fall of that year. A man fishing quietly off the end of Picnic Point received a terrible shock when a large snake-like head, with large jaws and blazing eyes, suddenly emerged from the water nearby. For a few moments the man was paralyzed with fear. Then, recovering his senses, he fled leaving his pole and fishing tackle behind. He related his experience to friends, but they just had a good laugh at his expense.

Not long after this incident, a university student and a girlfriend were sunbathing at the end of a frat house pier. They were lying on their stomachs with their feet toward the lake when the girl felt something tickling the sole of her foot. She glanced at her companion, thinking it might be him, but he was lying quietly with his eyes closed. She shrugged and lay down again, closing her eyes also. A few moments later the tickling on the soles of her feet came again. She quickly turned over and to her horror saw the head and neck of a huge snake-like creature extended above the surface of the water. Apparently it had been caressing the soles of her feet with its long tongue! Her cry aroused her companion and the two made a fast retreat to the nearby frat house.

The monster put in another appearance shortly thereafter in University Bay, while a couple were canoeing. Recognizing the creature from much publicized drawings of eastern states sea serpents, they hastily paddled for shore. After this incident other reports began to come in of sightings at different times and places in the lake. Fishermen were quick to report having seen it several years previously.

Skeptics thought the creature was merely a large pickerel, or gar fish with a collection of artificial baits clinging to its head, but witnesses firmly disagreed. The monster was named Bozho—probably an abbreviation of the name of the Old Indian hero-god, Winnebozho. When several water-spouts occurred in Lake Mendota, residents said, "Bozho was probably taking his shower bath."

As a rule Bozho was a good-natured monster only indulging in occasional pranks such as overturning canoes with his body or tail, chasing sailboats, uprooting a few lake piers and frightening bathers. When he finally disappeared, possibly by way of the Yahara River, people made more use of the lake.

Lake Monona, Wisconsin

Lake Monona, third in the chain of Madison lakes, had a prior record of a monster in its waters. The *Wisconsin State Journal*, June 12, 1897 printed the story, saying that the Monona sea serpent had made its appearance about two months earlier than usual that season. It was seen by several people in the vicinity of East Madison, the previous evening.

> They say it was at least twenty feet (6 m) long and traveled east on the surface of the lake. When Eugene Heath, agent of the Gear-Scott Company, fired two shots into it, the monster turned and came back; at this juncture either the monster or the spectators appear to have disappeared...
>
> Its appearance is not that of a serpent. Mr. Schott says, however, that he saw it plainly in the bright moonlight and its shape was like the bottom of a boat, but it was about twice as long. Mr. Schott's two sons saw it and were firmly convinced that it was a dangerous animal. When, soon after, two ladies desired to be rowed over to Lakeside neither of the Schotts, who had spent a large part of their lives on the lake, would venture out.

Some time later a monster, perhaps the same one, was also seen by different witnesses off the Tonywatha and Winnequah resort shore, on the east side of the lake. Years later, during dredging operations off the Olbricht Park shore, the pipes of the sand pump became clogged with some huge vertebrae that were believed to be from the skeleton of this sea serpent.

Lake Waubesa, Wisconsin

In the early 1920s a great snake-like creature was reported in Lake Waubesa, a large lake a few miles south of Madison. Concerned citizens wondered if the Lake Mendota monster had now migrated to Lake Waubesa.

One hot summer afternoon an Illinois resident of Edwards Park, on the east shore, rowed out on the lake to fish. He anchored his boat in good fishing grounds some distance from shore. He was surprised to see the quiet surface of the lake several hundred feet away begin to heave and move in swells. After a few moments, part of a huge body and then a large head broke the surface. The great creature seemed to be floating and sunning itself. The man watched and wondered what kind of giant eel or fish it could be. Then prudently raising anchor, he returned to shore. He estimated the creature to be sixty to seventy feet (18 to 21 m) long, dark green in color and with a serpent-like head.

Shortly thereafter, a couple who were swimming in the lake near their summer home on the Waubesa Beach shore, received a nasty shock when the head of an unfamiliar creature suddenly rose from the water a few feet away. Eyes glittering, it slowly moved closer. They wasted no time striking out for shore and the safety of their cottage.

From those and other reports the following summer, it seems the creature roamed all over the lake.

Mississippi River

The Mississippi River, the "father of waters," is a great drainage channel cleaving southward through the heart of the United States. Together with the Missouri, its principal tributary, the Mississippi drains an area of over a million square miles; nearly two-fifths of the United States.

No river as mighty as the Mississippi would be complete without its ancient monster. His origins lie deep in American Indian history. Temple mounds and isolated burials along the river contain carved stone effigies of this horned serpent. His likeness adorns pottery and is shaped in shell and wood. Belief in this monster is as keen in the Mississippi drainage as it is in the central valley of Mexico.

One of the earliest printed reports originated in the *Natchez Democrat* and was printed in the *Helena Independent* January 12, 1878. It seems the monster attacked a produce boat on the Mississippi. Some weeks previously the newspaper had published the particulars of a sea monster which, according to the captain of a towboat, attacked some barges his boat had in tow. The monster resembled an immense snake, with a bulldog head and a pelican bill about ten feet (3 m) long. "It lashed the water into foam with its tail, and spouted oblique streams of water forty feet high." Later, when the captain examined the barge, he found an ivory-like splinter from the creature's bill embedded in the timber.

At the time of publishing this report, the newspaper felt inclined to doubt the story, but after the second incident and subsequent interview with two gentlemen witnesses, the newspaper revised its opinion, "...we really think there is a big sea monster in the Mississippi River."

According to the two witnesses, on the night of January 9, they were floating down the river on Captain Ed Baker's produce boat. Near island number twenty-five, they were suddenly startled by a very loud splash in the water. Since they knew of the earlier report of a monster in the river they were very apprehensive. Some eighty yards (73 m) from the boat they detected a dark object swimming very fast towards the boat and making a lot of noise as it cut through the water.

> ...as it neared the boat it suddenly veered to the right, striking the stern oar and knocking it overboard. John Caughlin and Dud Kelley alone remained on the roof, the balance of the crew taking refuge in the cabin.
>
> The monster came near enough to enable these two gentlemen to get a full view of him. They judged him to be sixty-five feet (20 m) in length. His body was shaped like a snake, his tail forked like a fish and he had a bill like that of a pelican. His bill was fully six feet (2 m) in length. He had a long flowing black mane like a horse. When he swam his head was eight feet (215 m) above the water. It was a grand sight to see him move down the river.

The crew of the produce boat were so unnerved by the experience, upon reaching the landing, all except one man abandoned the boat. Captain Baker commented, "it is impossible to get a crew as the men think the monster is still following them."

Lake Minnetonka, Minnesota

On July 24, 1892, the *Chicago Sunday Tribune*, warned of a huge monster with the characteristics of a snake, toad and turtle "found to inhabit" Lake Minnetonka, a large lake near the center of the state of Minnesota. Thirty feet (9 m) long and with only "one green, blazing eye," the monster's appearances were said to be rare and occurring only at twilight.

> ...The few who have been at once privileged and horrified to behold this uncanny reptile describe it as of great length and of all the fantastic hues of Joseph's coat...
>
> The lower part of the body is shaped like a turtle. This portion is flat and nearly round being some ten feet (3 m) in diameter with a row of short stubby legs on each side, armed with sharp claws, like the turtle, alternating with other and much larger legs, which are found only among the batrachian species. Overlapping this turtle body some five feet (1.5 m) in front and twice that length behind writhe and twist the sinuous folds of the purely serpentine portion of this anomalous creature. It has no scales but is completely armored in a series of wart-like bunches as large as a man's head and varying color from white to black with all the intermediate shades of blue, purple, yellow and green.
>
> It wears no mane, but tufts of hair of all imaginable colors occur at frequent intervals on every side from head to tail and below the wide jaws appears a long, bushy goatee of coarsest fibre. Its jaws are furnished with broad pointed teeth and in and out between these rows of glistening bone plays the red, bifurcated tongue with amazing swiftness.
>
> The monster's mode of propulsion is a curious cross between the ordinary snakelike natation and the awkward paddle of the turtle, but its speed is remarkable.
>
> When suddenly and without premonition, this most horrible hybrid rises to the surface its vocal organs emit a peculiar noise, half-roar and half-scream and hiss which, added to the tumultuous lashing of the water with the restless tail, produces a most startling effect upon the ear.

While this report is admirable for its attention to detail, it is hard to imagine any monster staying still long enough to allow such a graphic description—unless it was a dead monster! As this clearly was not the case, perhaps the author became a little confused between myth and reality, for there is a legend attached to Lake Minnetonka.

More than 100 years ago, the legend says, Edmond Dornier,

an old French settler, lived on Crane Island in Lake Minnetonka. One day while his lovely nineteen-year-old daughter was fishing off the island, her canoe suddenly upset throwing her into the water. She would surely have drowned but for the swift action of Mehawanta, an old Sioux Indian Chief, who swam out and brought her to shore. Tenderly laying her on the shore, Mehawanta began resuscitation.

Meanwhile Dornier, who was 100 yards (91 m) away and did not see the mishap, suddenly saw the Indian bending over his daughter. Misinterpreting the situation, he lifted his rifle and shot the old chief dead.

So terrible was the father's despair when he discovered his awful mistake, that for the rest of his life he was haunted by the most poignant grief and remorse. Legend says the great serpent turtle of Minnetonka is the material embodiment of Donnier's remorse.

Paint River, Michigan

While some monsters would seem to be longtime residents of certain lakes and rivers, others appear only once or twice and then disappear. The creature in the Paint River, Iron County, in the center of the upper peninsula of Michigan, appears to be one of the latter.

The monster was seen in 1922 by an unidentified lady and a Swedish woman named Johnson:

> I was walking down the hill toward the river...Mrs. Johnson was walking up the hill and we met on the knoll about halfway, which was very near the river. We got a very good look at the animal...both saw it at the same time, and stood stunned, speechless, watching it till it went out of sight. Mrs. Johnson...asked me in Swedish, "Did you see what I saw?" I assured her that I did. She went on...very excited, saying, "It had a head bigger than a pail." She then made me walk back up the hill to our house, and I had to verify everything she said to my mother.
>
> ...the head stood straight above the water; the body was dark color; the body did not move like that of a snake, but in an undulating motion. We could see humps sticking out of the water, and I recall counting six of them...It was swimming

> north up the river between two bridges...this distance could be the length of a city block, and this monster must have been clearly half of that distance (length). It submerged under the bridge, but we followed its wake on up the river.

Mrs. Johnson verified the other lady's description of the creature. According to the late, internationally known zoologist, Ivan T. Sanderson, *Saga Magazine*, the significance of the report was the undulating motion, which is typical of all the long necked sea serpents both in the sea and freshwater. Snakes move with a side to side motion; they cannot undulate. This is the case with all reptiles. Mammals, however, such as whales and seals, do "hump" along.

Walgren Lake (Alkali Lake), Nebraska

Walgren Lake, near Hay Springs, has been the home of an elusive monster for a long time. In the fall of 1922, J. A. Johnson and two friends were camped a short distance from the lake. Having risen early to be ready for duck hunting they started walking around the lake close to the shore in order to jump any birds. Suddenly, as they rounded a slight rise in the ground, they came upon a strange animal standing in the shallow water, less than twenty yards away near the shore.

> The animal was probably forty feet (12 m) long, including the tail and head, when raised in alarm as when he saw us. In general appearance, the animal was not unlike an alligator, except that the head was stubbier and there seemed to be a projection like a horn between the eyes and nostrils. The animal was built much more heavily throughout than an alligator. Its color seemed a dull gray or brown...
>
> There was a very distinctive and somewhat unpleasant odor noticeable for several moments after the beast had vanished into the water. We stood for several minutes after it had gone, hardly knowing what to do or say, when we noticed several hundred feet out from the shore a considerable commotion in the water.
>
> Sure enough the animal came to the surface, floated there a moment...lashed the water with its tail, suddenly dived and we saw no more of him.

Johnson's description of the creature, as "dull gray or

brown" in color and with a "projection like a horn between the eyes and nostrils," is reminiscent of the one given by Ernest Denks of the White River monster and could suggest that Walgren Lake and the White River harbor the same (gigantic penguin) type of monsters. Johnson further stated that he could name a number of other persons who had also seen the creature.

Big Alkali Lake, Nebraska

The following year, *Omaha World-Herald*, July 14, 1923 printed another report from the area. It seems a monster gave several Nebraska and Texas men a scare. On the previous Monday evening, George Locke of Central City, Bob Cook of Lakeside, and two Texans started to drive overland from Lakeside to Hay Springs. It was after dark when the men arrived at the shore of Big Alkali Lake, 138 miles northeast of Lakeside and their car became mired in the swampy ground at the edge of the lake. They tried for two hours to release their vehicle, but were finally forced to admit defeat. Settling themselves in the car, they decided to wait until morning when they could get help.

Around 2:00 A.M., Locke was awakened by a commotion in the water. He was almost startled out of his wits to see a head and horns of an enormous creature emerge from the lake and move towards the car. With an unearthly yell, Locke awakened his companions. All but Bob Cook, the driver of the car, fled. Cook stayed where he was until the fog from the monster's nostrils enveloped the car, then he too followed in his companions' wake and made for the nearest farmhouse two miles (3 km) away.

Next morning, when they returned to the scene, they were amazed to find the car still intact. With the aid of a team of farm horses, the car was finally pulled from the mud.

On the strength of the report, the Alliance Anglers Club sent an order to a Boston whaling ship outfitting company for a harpoon, line and whaling gun. It was their belief the monster could be captured. A posse was organized and upon receipt of the outfit, would make a "desperate attempt to rid the waters of the animal." Bob Cook, who could identify the monster on sight, was to head the party.

The enthusiastic monster hunters, however, had not reckoned on the reaction of the property owners around the lake.

They viewed the whole venture with disdain. They did not like the idea of people camping out and tramping all over their property. They announced they would not allow the lake to be disturbed and constructed a fence around its limits. The whole monster hunt dissolved in a dispute over leasing of the lake and in ridicule from outside.

It is interesting to note, however, that elsewhere on the Great Plains certain pools have been identified where creatures such as Bob Cook and companions describe, are supposed to live. According to *Pursuit*, April, 1973, "One anthropologist Stanley Vestal, has recorded that what are known simply as 'water monsters' have a reptilian form and horns like a buffalo. They are seen in pools, sinkholes, and rivers."

Late-comers to the Plains also encountered these creatures. In 1934, one farmer in Brookings County, South Dakota was forced to take to a ditch on his tractor when of one of these giants crossed his path. The animal's track was later followed until it disappeared into Lake Campbell.

Leech Lake, Minnesota

An article in the *Minneapolis Star*, October 1, 1976 about Leech Lake, a big northern lake in Minnesota, asked: "Might there not be monsters swimming in those waters?" Apparently, at Walker a few days earlier, something strange had appeared on graph-type fish finders. John Aldrich and Richard "Skip" Christman of Vexilar, Incorporated, manufacturers of the fish finders, were curious. They had spent two days at Leech Lake equipped with heavy downrigger equipment, the kind usually used on Lake Superior or Lake Michigan.

They saw the huge targets on the graph, suspended at approximately sixty feet (18 m) in water at a depth of 90 to 150 feet (up to 46 m). Live bait and artificial lures were used. Large fish and numerous bait fish just scattered when the lures approached.

The men's only catch the first day was a three pound (1.3 kg) northern pike, on a downrigger set forty-eight feet (15 m) in fifty-five feet of water, with a small gold spoon as a lure. The second day they caught a three pound walleye at the same depth.

The graph recording continued to show the large targets,

also schools of smaller fish, yet nothing else was caught. They tried the same area at night. No large suspended fish were charted in the same depth, but a horizontal bank of fish appeared on the screen; indicating very large fishes.

What the big ones were still remains a mystery to the two men, as do all the other strange manifestations in rivers and lakes of the midwestern states region.

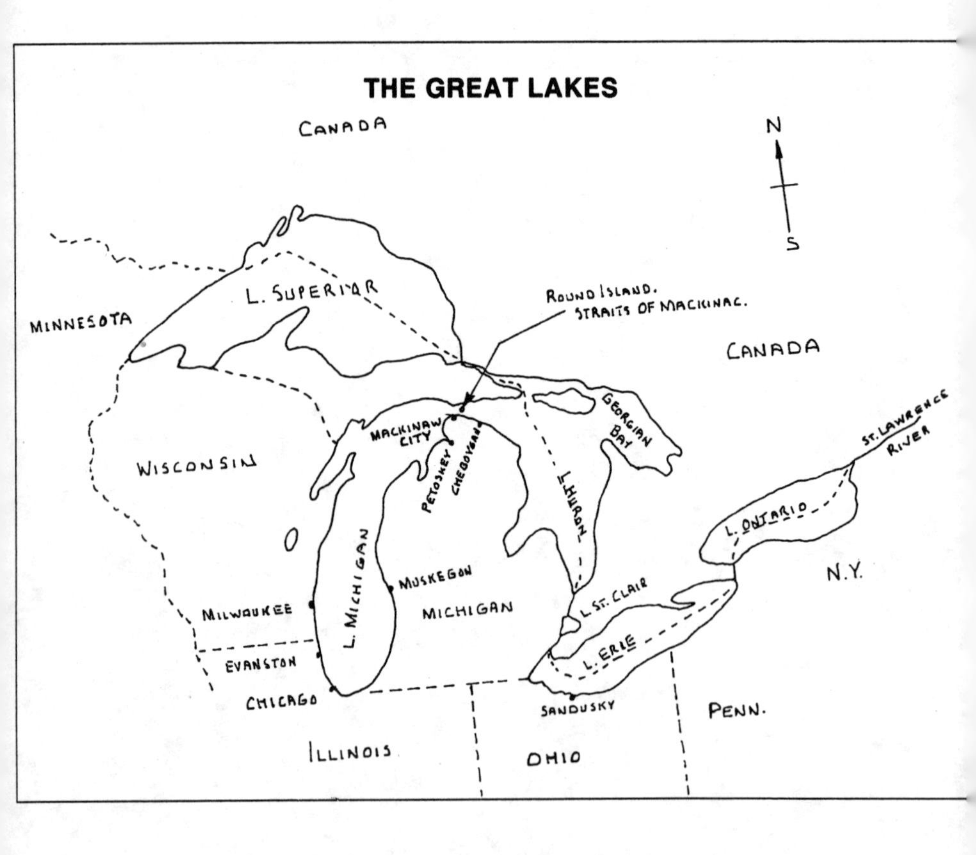
THE GREAT LAKES
CANADA
N
S
L. SUPERIOR
MINNESOTA
ROUND ISLAND.
STRAITS OF MACKINAC.
CANADA
GEORGIAN BAY
ST. LAWRENCE RIVER
MACKINAW CITY
PETOSKEY
CHEBOYGAN
WISCONSIN
L. HURON
L. ONTARIO
N.Y.
MUSKEGON
L. MICHIGAN
MICHIGAN
MILWAUKEE
L. ST. CLAIR
L. ERIE
EVANSTON
CHICAGO
SANDUSKY
PENN.
ILLINOIS
OHIO

8

"T'was Black And Oily At Mackinac"

Lake And River Monsters The Great Lakes: Lake Michigan; Lake Erie; Lake Huron

With lake monster reports rife in the Midwestern region, one could hardly expect the Great Lakes to be without their watery residents. A motley crew seem to dwell in the depths of Lake Michigan. The largest lake in the United States and third largest of the Great Lakes, Lake Michigan is the only one entirely in the United States. A wealth of reports have come from this lake—several from the Illinois coast.

On August 7, 1867 the *Chicago Tribune* declared: "that Lake Michigan is inhabited by a vast monster, part fish and part serpent no longer admits of doubt."

Shortly thereafter, the crews of the tug *George W. Wood* and the propeller *Skylark,* saw the creature lashing through the waves off Evanson. It was between forty and fifty feet (12 and 15 m) long, with a neck as thick as a human being's and a body as thick as a barrel. On the morning of August 6, fisherman Joseph Muhlke met the same, or similar, creature on the lake a mile and a half (2.4 km) from the Hyde Park section of Chicago. After that, reports seemed to peter out.

In 1881, much excitement surrounded the capture of a "big sea serpent" in Sandusky Bay, Lake Erie, the most southerly and fourth largest of the Great Lakes, shared by Canada and the United States. According to the *New York Times*, fishermen Clifford Wilson and Francis Cogenstose of Cincinnati, were fishing in the lake when a large marine animal measuring twenty-five feet (7.6 m) long and about twelve inches (30 cm) through the thickest part, rose from the water beside their boat.

Somewhat unnerved, they whacked it over the head with an oar, knocking it senseless. Then they fastened a line to its head and towed it to shore. The creature began to revive when they reached shore. The two men obtained a packing box six feet (2 m) long, three feet (1 m) wide and approximately two feet (61 cm) deep, coiled the creature into it and nailed it tightly shut.

By that time, the *Times* said the monster was thrashing about dangerously. Neither the "owners" nor any of the scores of anxious onlookers would take a chance on opening the box to show the "serpent" to skeptics.

Police Captain Leo Schively, E. L. Ways, managing editor of a local afternoon paper, C. J. Irwin and Mel Harmon, of a Sandusky morning paper, did see the serpent while it was being boxed up. They joined with the two fishermen in describing it as "a huge, snakelike beast, colored black, dark green and white and having a hide resembling that of an alligator."

The newspaper said the "owners" were undecided what to do with the creature. Apparently, the capture followed reports by a number of people of a sea serpent in the lake.

Some years later another report appeared in the *Chicago Sunday Tribune*, July 24, 1892. While crossing the lake, Captain Jenkins of the steambarge *Fenton*, sighted what at first appeared to be a wreck of some kind. Upon moving closer they discovered it was not a wreck at all, but a huge serpent about thirty feet (9 m) or more long.

> The tail of the monster was laterally compressed thus adapting it to the same purpose in locomotion through the water as the caudal fin in fishes. The head was nearly a foot (30.5 cm) in length, the nostrils being placed not as in ordinary serpents at the end of the snout, but above and the eyes, blazing like

two balls of fire, were about two and a half inches (6.4 cm) in diameter. The neck was very short and thick set and the mouth, turned upward instead of downward, was of huge cavernous dimensions when the animal opened its jaws so as to display its forked tongue. The color was black with yellowish white bands on the body and white patches upon the head.

On passing the huge reptile it was observed to rear its head and neck out of the water and fall into the wake of the boat as if in pursuit of prey. For twenty or more miles (32.2 km) the chase continued, the serpent equalling the speed of the steamer and swimming gracefully most of the time with its head and neck only out of the water, but occasionally rearing upon its abdomen so that it seemed to stand up straight above the water for about fifteen feet (5 m), as if to take a survey of the deck and see if there was any prey there worth seizing. At intervals, the animal approached the ship closely and rearing itself, as it were, on its haunches seemed disposed to dart on board and seize some of the persons on deck. At such times the experience was thrilling in the extreme.

Nevertheless, it gave the opportunity to observe the situation of the serpent's nostrils, the conical-shaped teeth pointing backwards in the open mouth and reddish color of the abdomen.

Finally, after about twenty miles (32 km) were thus passed, the huge monster as if wearied, or deeming further pursuit hopeless, abandoned the chase and swam gracefully away.

Around the same time, reports of a monster came from the Michigan coast of Lake Huron, the second largest of the Great Lakes, shared by Canada and the United States. This lake receives the waters of Lake Superior by St. Mary's River and those of Lake Michigan by the Straits of Mackinac. Its outlet is through the St. Clair River and Lake and the Detroit River to Lake Erie. The *Chicago Sunday Tribune*, July 24, 1892 carried several reports of the monster.

"T'was Black and Oily at Mackinac," read the headline to a special report from Mackinac Island, dated July 23. J. Frederick Stevenson made a statement to the effect that while he and some friends were swimming near Round Island they had a disturbing encounter with the monster.

> ...They had several times gone (bathing) this year, as today in a Mackinaw boat. They left the water hurriedly in fright caused by the appearance of some kind of monster in the form and general appearance of a huge snake. The creature did not attack them or come toward them...but it came near enough to one of the party, who was at some distance from land, to seem so. It was taking its own course, which seemed to be in a southeasterly direction, towards the mainland and without apparently noticing the presence of the bathers.

Mr. Stevenson further stated he was acquainted with sea serpent lore, having often heard them described. The creature, whatever it was, did not possess the traditional fiery-eyed dragon head, nor did it have fins or any coloring as far as he could see; but was black and oily and made a singular whirring noise as it passed. Except for its enormous size, the creature seemed similar to a snake in appearance and in the manner it traveled. The most striking and startling thing was the peculiar noise it made, sounding like the singing buzz of machinery.

Another report came from Petoskey, on Lake Michigan. A party of tourists were out sailing on the lake about twelve miles (19.3 km) from Petoskey, when one of the party saw what appeared to be a huge mass of tumbling waves. He called it to the attention of his companions. Upon coming closer they saw a monstrous sea serpent sporting in the water. It would lash the water into foam for a great distance, roll and tumble about and then lie perfectly still. At this point its shape and size could be clearly seen. At times it would dive and disappear only to reappear a short distance away. At other times it would raise itself almost entirely out of the water, when its ugly looking fins were plainly visible.

Finally the monster gave three or four terrible lashes with its immense tail and then sank out of sight. Estimated length of the creature was from sixty to seventy-five feet (18 to 23 m) and the body 4 feet (1 m) in diameter. Its head was clearly seen and its vicious-looking eyes were as large as dinner plates. The jaws were immense and fairly bristled with ugly sharp teeth. The color was a dark brown, which lightened towards the tail.

The body tapered like a snake's, but was not very long for its size.

"The water was quite calm, and there was but little wind, and there can be no doubt that the people saw what they claim," the report concluded.

It began to look as though the same monster seen earlier near Round Island, Lake Huron, and before that in Lake Erie, might be traveling between the lakes. Shortly thereafter, an astonishing story came from Muskegon, further down the Michigan coast.

It seems the schooner *Cheney Ames* had been mysteriously dashed against the sharp corner of the south pier at the harbor, receiving a rent in her hull. Prompt action by Ed Maloney, one of her crew, kept her afloat until she finally went down in shallow water several hundred feet up the channel.

When the crash occurred, realizing the danger of sinking, Maloney sprang overboard. He succeeded in placing a tarpaulin over the gash; the pressure of the water holding it in place. His prompt action kept the *Cheney Ames* afloat, but Maloney was hurled by the heavy seas aft and underneath the vessel.

Since he was reputed to swim and dive "like an Alaskan seal," the crew had no fears for his safety. He passed underneath the vessel, coming up to the surface abaft the stern on the opposite side. But instead of immediately striking out for shore, as was expected, he seemed stunned or "paralyzed." He drifted to the pier where he was hauled from the water in a state of shock. His face was ashen. As he recovered, he told an incredible story, yet one that tallied with current stories of the area.

As Maloney was passing under the vessel he glanced through the water as one would through open air. What he saw "froze his blood and dried up his fountains of speech."

> Lodged between the wheel and rudder of the peculiarly constructed schooner, and coiled...several...times about the rudder was a serpent fully sixty feet (18 m) in length. Its jaws, in which gleamed ferocious fangs, were distended and the gleaming bead-like eyes of the beast looked earnestly into the eyes of the sailor as he passed.
>
> The lower part of the body was yellowish pink and white, while the upper portion of the reptile was glossy black. The

wheel had cut a gash in its side from which the blood spurted and discolored the water in the wake of the vessel.

Before the man had revived sufficiently to tell his story, the monster had freed itself from the rudder and wheel and plunged away into the depths.

Freed of its burden, the rudder suddenly responded to the desperate efforts of the wheelsman. The disabled schooner was thrown to the north side of the channel and shortly thereafter sank in shallow water.

Further investigations revealed that around two o'clock, approximately one hour before the wrecking of the vessel, the monster had been seen by William Dixon, Superintendent of the Water Department. It was floating in toward the harbor in a spiraling motion, like a "huge, inverted corkscrew. Then it changed its position to a mammoth, yet graceful interrogation point." Mr. Dixon had made his report before the ill-fated *Cheney Ames* encountered the monster.

The *Tribune* concluded by saying that the presence of this mighty reptile had caused some nervousness on the part of summer tourists at Lake Harbor and the coast was constantly being patrolled in hopes of catching a glimpse of the "terror of the lake."

On July 24, the *Tribune* also reported another sighting of the monster. While Captain McKee, of the two-mile crib, was sitting on a guard rail by the lake watching the sunrise, something strange caught his eye.

At first he thought it was a big wave, not white crested or breaking, but a solid wall of water moving slowly in the direction of the crib. About 500 feet (152 m) away, the wave seemed to disappear only to reappear again in a circular form. The captain ran to the office and secured a marine glass, which he focused on the object. He could not believe it was a freak of wind and water, neither did he believe stories of sea serpents. What he saw through the glass shattered his skepticism to pieces.

There, lashing its long tail about in the water, was a serpent fully 180 feet (55 m) long. The monster came to within 200 feet (61 m) of the crib, raised its head some thirty feet (9 m) from the water and slowly surveyed its surrounding. Its head was like a gila monster's, flat and broader at the gills than at the snout.

It was yellow and black, with rather small, pinkish eyes that seemed out of proportion with its immense size. Horns, three feet long and the color of dirty ivory, protruded from its forehead. From the ugly mouth hung a crop of reddish whiskers. The body appeared to be scaly. The powerful tail was forked.

Evidently satisfied with its surrounding, the serpent turned and disappeared beneath the surface. Odd sightings of this monster continued to come in at intervals. In the 1900s, it was reported from the Jones Island shore of Lake Michigan, in the state of Wisconsin.

While some market fishermen were setting their nets one day, the large head of a ferocious looking beast appeared above the surface of the water. The fishermen who were close by obtained a good look at the creature before it submerged. Upon returning to shore their story was met with ridicule. But shortly thereafter, while some young men were sailing a catboat in Milwaukee Bay, they saw what looked like a large cask floating some distance from their boat. Closer inspection revealed it was the head of a large serpentine animal which appeared to be floating at rest. Bettering their curiosity, they quickly changed course.

This same serpentine creature was seen by a man early one morning in the Milwaukee River. He was leaning on the rail of the Michigan Street bridge when he saw a large grayish green body moving down the river in the murky water. Suddenly it dived and disappeared. He told the bridge tender what he had seen, but the warden said it was probably just a log or large timber. Later the same morning, another report came from further down the river near its mouth. Random reports of the monster continued for quite some time.

Eighty years later, a report in the *Detroit News*, June 25, 1976 suggested the monster might still be alive and well and on a jaunt to Lake Huron again. A motel owner at the northern Michigan resort town of Cheboygan reported something that looked like "a big snake, eel, or serpent, or monster" off the Lake Huron shoreline half-way to Mackinaw City. Sheriff Stanley McKervey was studying the Monday sighting, together with up to a dozen reports of "other mysterious things" since seen on the lake. The reports told of "snakes—maybe twenty-five feet long (8 m), with the head slightly out of the water and moving about three miles (4.8 km) an hour."

Large crowds gathered along thc shores. McKervey watched the creature, or creatures, through field glasses for about fifteen minutes. A fish biologist from the State Department of Natural Resources said they may have been eels that slipped in from the ocean along the St. Lawrence Seaway.

It is very probable that in the early days of North America eels of enormous proportions did exist in the Great Lakes. Increased shipping accidents, such as that of the *Cheney Ames* and pollution could have contributed to their disappearance—at least, from some of the Great Lakes. A handful of sightings in recent years suggest that Lake Erie's serpent might still be around.

Summer 1990 produced a rash of sightings of a blackish creature forty to fifty feet (12 to 15 m) long, with bulging frog-like eyes. "The head was probably twice the size of a

SIGHTING NEAR PORT DOVER

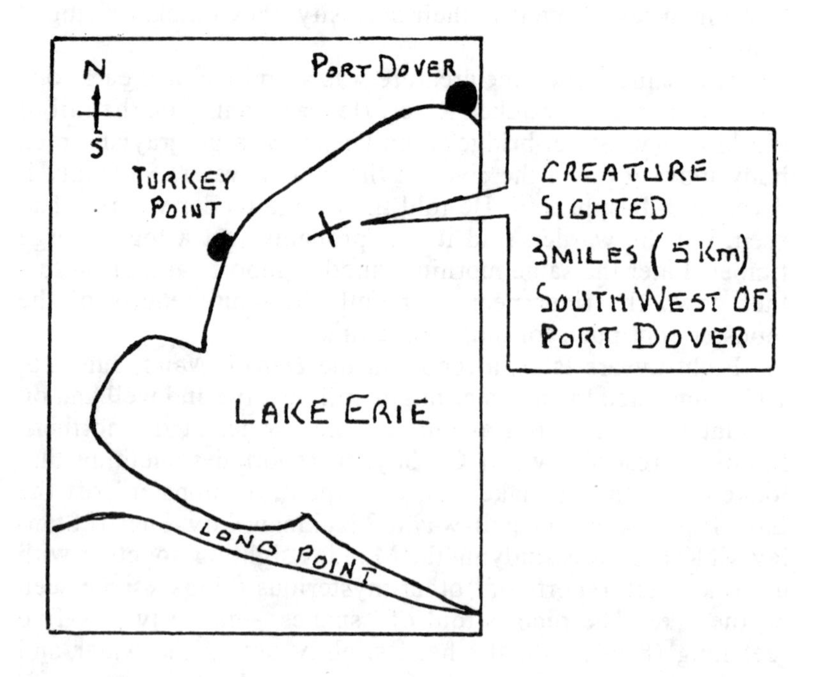

football," said college teacher, Mary Landoll, of Huron, Ohio, who watched the monster from the shore of her summer property on Lake Erie. Another witness, Martha Iskyan, a University of Toledo counsellor, saw the creature on July 14, while she was out sailing. "It started swimming with snake-like movements," she said. "It would make an S-shape, move forward, straighten out, make another S-shape, and so on. Then I noticed two small horns sticking up from what I took to be its head."

There was another sighting on September 4 off Cedar Point—in fact, so many people have seen the Lake Erie monster, that an Ohio newspaper set up a hot-line for people to call if they spot the mysterious creature.

On October 8, 1994 the Hamilton *Spectator* reported a new sighting—this time, from the Canadian side of the lake, near Port Dover. According to two coastguards who witnessed the event, the creature was large, smooth, and dark colored. It moved very fast, was visible for about ten to twenty seconds, and dived quickly, quietly and deep when the pair turned their boat towards it.

Whatever the two men saw, the creature's identity will remain a mystery. Perhaps the persistent monster who pursued the steambarge *Fenton* on Lake Erie, in 1892, or one of its descendants, is still alive and well!

9

"Slimy Slim" And Other Long Necks

Lake And River Monsters Rocky Mountain States: Idaho; Montana; Utah

Fewer people live in the Rocky Mountain states than in the state of Michigan, but the region occupies more than a fourth of the United States. It is an area of awe inspiring scenery where glaciers and rushing rivers have gouged deep valleys and canyons, such as Yellowstone Canyon in Yellowstone National Park. The churning waters of Idaho's mighty Snake River rush through Hell's Canyon, which is deeper than the Grand Canyon and plunge down rugged cliffs at Shoshone Falls from a height greater than that of Niagara Falls.

Idaho, in the northwestern corner of the region, still has many areas that remain largely a mystery. It has more than 2,000 well-known lakes. No one knows how many hundreds of lakes in the state have not been discovered. Many mountain lakes have been seen only from the air.

What better setting for monsters! Many are the wonderful tales that have come out of this area. One of the earliest stories was told by an old-timer in Swan Valley, southeastern Idaho and was printed in *IDAHO LORE*, Federal Writers' program 1939. It concerned the mighty Snake River:

> On August 22, 1868, I wuz crossin the river at Olds Ferry when I seen somethin stickin outa the water. It pears tuh be alive, fur it don't jist drift with the current but keeps a-movin this way and that, an they's big ripples all around it. Purty soon it comes nearer and I see somethin like a elephant's trunk shootin up, an the dang thing starts tuh spoutin water—then somethin comes tuh the surface that looks like the head uv a snake, but it's ez big ez a washtub, only flat-like an hez that gol-darned horn a-stickin up out uv it an hez long, black whiskers at the sides uv its face.
>
> By that time I wuz on the other side so I whips up my horse an rides up the side uv the hill where I cud git a good sight uv it. I gits off my horse and ties him tuh a quakin asp behind some brush an then I watches the thing. It's swimming towards the bank. Purty soon it gits there an heaves itself outa the water an starts slitherin up the hill in my direction—an say, yuh never smelt sech a stink in all your life! My horse starts a-snorting and a-raring, and breaks loose and lights out through the timber. An all the time that stinkin reptile is gittin closter tuh me. I make out it's about twenty feet (6 m) long, and outside uv that horn that keeps shootin up an down out uv its head, an the whiskers, an a pair uv big fins, er wings, er whatever they is, all shiny-like in the sun, that comes outa the sides uv its neck, it looks down tuh the middle like a big snake, ez big around ez a calf—uv a kinda greenish-yaller color with red an black spots on it. An then all uv a suddint right then it pears tuh turn intuh a fish, with scales ez big ez yer hand, all colors uv the rainbow, shinin like big pieces uv colored glass in the sun—and then all the shinin peters out an all the rest uv it is jist a grayish colored horny tail, like a big lizard's er mebbe a crocodile's—an all the way from where that doggone horn kep shooting up and down on its head clear down tuh the tail they's a line uv shiny black spines with a hook on the end uv em like a porcupine's.
>
> It keeps a-coming closter and closter to where I'm a-standin with my rifle cocked an the closter it come, the worst it stunk.

I'm feelin purty sick but I sez tuh myself no sech critter ez that oughta be let tuh live. An I gits a draw on it and lets it have a slug in the eye. It gives a beller an raises up on its tail twenty feet (6 m) high and starts fer me a-hissing and a-snarlin. Its mouth is open a foot (30 cm) wide showin fangs ten inches (25 cm) long an a red forked tongue that keeps a-dartin in an out a-spurtin green pizen. I let's it have another slug in its yaller belly—and it drops an thrashes around on the ground a-hissin an a-snarlin and a-bellerin somethin awful, tearin up the earth, an knockin down brush and trees, an smashing everythin around it.

Bye an bye it's still, and I goes tuh take a look. It's a-layin on its back and I see it's got twelve short legs a-growing out uv its belly; the first pair, next tuh the tail, hez hoofs on em, the next pair hez long claws, and the last below the fins, er wings, er whatever they is, is hoofs agin. An everywhere I see, they's black patches where it's spit pizen, an everythin it's tetched is a-witherin an a-dyin—trees and bushes an grass an everythin.

Well, I goes tuh git me some kind uv a rig tuh take the critter tuh town in. I calculate tuh have it stuffed an show it at the fair fer so much a head, an mebbe make a little money off'n it. I gits a team an a dead-ax wagon and six fellers tuh help me haul it down tuh the wagon an load it, fer we can't drive right up tuh where it is on accounts the timber. I tell em tuh wear gloves and I takes a old tarp along tuh wrap around the speciment so's when we handle it we won't non uv us git pizened.

When we're a hundred rods (503 m) er so from the place, we begins tuh smell that stink agin and the horses ack skeered. I'm a-drivin and I try tuh quiet em down but they keep on a-snortin an a-rarin an two uv the other fellers has tuh git out an hold em down but we can't do nothing with em so I got tuh back a ways till I find a place tuh turn and then I turns round and heads em the other way. We leave one feller there with the team an the rest uv us goes on.

It tain't no pleasant trip fer the nearer we git the worst the atmosphere gits. I don't know how I looked but the other

> fellas hez ahold their noses an wuz lookin purty pale around the gills. One feller got sick and didn't come no further.
>
> When we gits there we see where the ground is all tore up, an the brush all trampled down fer fifty foot (15 m) and quaking asps an cottonwoods knocked down an a-layin on the ground—an everthin the pizen hit wuz dead. But nary sight uv the critter. But we finds its trail along the flattened grass an busted brush, and it leads down an smack intuh the river.
>
> We waited round fer the rest uv the day an half the town wuz there with us watchin the river, fer they's heard about the strange critter; but nobody sees hide ner hair uv it, ner spout ner tail.
>
> They say a snake don't die till sundown. I dunno what the durn thing wuz, but mebbe it went down in the water tuh die—or mebbe it *didn't* die.
>
> Boy an man, Ive hunted an trapped an fished all over the state fer nigh ontuh seventy-five year, I've ketched some purty queer fish by hook and trap, but I ain't never seen nothin tuh compare with that speciment.

What the old-timer actually encountered remains a mystery, but almost anything seems possible in mysterious Idaho!

Payette Lake, Idaho

West of Swan Valley and still connected with the mighty Snake River, lies Payette Lake. This seven mile (11.3 km) stretch of deep, blue, cold water is a remnant of the glacial age. At an altitude of over 5,000 feet (1,500 m), the lake is surrounded by mountains. Fed by mountain springs and streams, it has an outlet into the north fork of the Payette River, an important branch of the Snake River. The ridges circling the lake are pine covered; the slopes below are dotted with weekend cottages around the lakeshore. Beef cattle graze in a western-story valley. For well over fifty years, Payette Lake has been the home of "Sharlie," alias "Slimy Slim"; a shy monster who shows himself only rarely and usually after dusk.

In the 1930s most reports came from workers at the Brown Tie and Lumber Mill, which later became the Boise-Cascade operation. But in the summer of 1944, some thirty people, mostly boaters on the lake, witnessed his periscope-like head above the blue surface. At first witnesses discussed the serpent only in private, among close friends, but when Thomas L. Rogers, auditor of a Boise firm had his experience he decided to speak up.

Time Magazine, August 21, 1944, printed his description of the monster:

> The serpent was about fifty feet (15 m) away and going five miles (8 km) an hour with a sort of undulating motion...His head, which resembles that of a snub-nosed crocodile, was eighteen inches (46 cm) above the water. I'd say he was thirty-five feet (11 m) long.

Idaho's previous skepticism changed overnight. The monster was nicknamed Slimy Slim. Eager fishermen tried to catch him with deep-sea tackle and photographers descended upon the lake with cameras hoping for a glimpse of their shy monster. Many were the theories as to his origin. *The Idaho Sunday Statesman* took a more facetious attitude towards him. They felt the following story was as good an explanation as any:

One hot summer's day while Paul Bunyan was fishing the Snake River, he tied the shore-end of his sturgeon-line to his huge, blue ox named Babe. Shortly thereafter, a big horsefly came along and bit poor Babe at the exact moment a sturgeon took the bait. Babe twitched so violently from her wound that the huge fish went sailing all the way to Payette Lake!

Time Magazine commented on the story: "A jerk like that could well have given the creature a curvature of the spine," (Slimy Slim is purported to have three humps on his back). "And then Slim developed his periscope neck by nostalgically trying to peer back over the hills toward the scenes of his childhood."

Other explanations of the strange presence in the lake were more serious. Most popular was the theory that a huge sturgeon was trapped in the lake.

During the 1950s, after another spat of sightings, the *Payette Lake Star* sponsored a contest to officially name the creature. Hundreds of names were submitted, some from quite widespread areas. Mrs. Lee Isle Hennefer Tury of Springfield, Virginia, formerly of Twin Falls, wrote: "Vas you dere, Sharlie? Name him Sharlie." She won the contest and the monster was officially named.

One of the best sightings was made by Pauline Miller, a McCall resident for over forty years. Seated at her desk in her office in the Lakeview Hotel, directly across from the lake, she noticed an object which she took to be a large boat, floating near the city dock. Since it was not usual for such big ships to be on the lake, she left her office and went across the street to get a better look.

Mrs. Miller, the *Idaho Statesman* reports, discovered with a shock that the boat was really "a big fish of some sort." She stood for two or three minutes watching the creature and was able to give a good description.

Its head was under the water as if it were feeding, but its back which was sticking up, "looked like wet rubber...black and slick." There were no scales on its back, but there were "three or four indentations like ditches, and as it swam, the water ran through these dents and over the back."

The monster suddenly turned in the water, revealing a distinct line below what she now saw was a shell-like covering over its back. The flesh underneath was "brilliant white, with huge scales that glistened in the sun...like mother of pearl." The tail part of the creature was submerged, but the body tapered from approximately "as wide as a car length," to a thickness of about three feet (1 m) at the rear.

Other sightings of Sharlie occurred in 1953. Seven people all gave descriptions much the same. It resembled a snake, was raised about eighteen to twenty inches (46 to 51 cm) above the waves; looked much like a periscope and appeared to have four humps on its back. The tail and fins resembled an airplane rudder.

Sharlie is a kindly monster. He keeps much to himself and has never been a real menace to swimmers or fishermen. There has been an absence of reports for a number of years now. But Payette Lake is more than 3,900 feet (1,190 m) deep in places. There are many vast cavern shelves where Sharlie could be hiding. This is the theory put forward by *Incredible Idaho* magazine. One of these cavern shelves may extend back for miles, which could account for Sharlie's infrequent appearances. Another theory is that a secret underground spring warms his habitat and that is why he is seldom seen in cool weather.

Sharlie is not the only monster attributed to Payette Lake.

In an article in *Saga Magazine*, entitled: "'Monster' Hunting," zoologist, the late Ivan T. Sanderson describes these creatures as much like other long necks (plesiosaur types) "with heads variously described as looking like those of cows on long, slender necks, and with humps behind, and proceeding by an up-and-down motion, making a big bow wave at speed." They are yellowish in color. Local newspapers have reported sightings by dozens of people at the same time.

Flathead Lake, Montana

Flathead Lake lies near the long, thin portion of Idaho, on the western edge of Montana. It is the greatest natural freshwater lake between the Mississippi River and the Pacific Ocean and vies with Lake Okanagan in Canada and the famed Loch Ness in Scotland.

Flathead Lake was formed millions of years ago when gigantic glaciers pushed down from the north pole regions, filling the valleys and pushing their way to the south—ever gouging out the earth. Eons ago this lake was an enormous body of glacial water close to 1,000 feet (300 m) deeper than it is today. It covered nearly all the central part of western Montana. Only the higher hills and mountains showed their heads, like islands above the water.

Today, the lake is only a fraction of its ancient size; thirty-eight miles long (61 km) fifteen miles (24 km) at its widest points and nearly 350 feet (107 m) deep. It has an area of 188 miles (302.5 km) and 123 miles (198 km) of beautiful shorelines.

It seems only right such magnificent clear, cold depths should house several monsters. Ivan Sanderson described them as long necks. They have been reported almost every year for decades. The reports are not casual ones. Often, as in the following case, there are numbers of witnesses.

On Saturday, June 15, 1963 twelve people saw off Finley Point what appeared to be a large log floating toward the main part of the lake. At first they thought it was being towed by a boat, but when the object began to move in an undulating motion, they realized it was an animal they could not explain.

Three years earlier, in September 1960, Mr. and Mrs. Gilbert Zigler had seen a monster in the lake at close quarters. Mr.

Zigler was grounds keeper at thc Polson Country Club, during the summer. The rest of the year he was a city policeman and part-time deputy sheriff.

About seven-thirty one Friday night, the Ziglers heard waves crashing against the shore of the lake. They went outside but found no boats were churning the lake. High waves, splashing the shore, gave the impression that a large animal in the water was scratching its back against the end of the pier.

Mr. Zigler turned back to the house for a rifle. Mrs. Zigler went closer to the pier. At that moment the monster raised its head out of the water. The head was about the size of a horse's and there was "about a foot (30 cm) of neck showing. It was a horrible looking thing," Mrs. Zigler said. Her scream brought her husband running, rifle in hand. He arrived to see the creature swimming east parallel to the shore, at a very fast speed and making a very large wave.

Paul Fugleburg, editor of the *Flathead Courier*, who kept a dossier of reports, offered a reward for the first photograph of the Flathead monster. So far no one has claimed it.

Attempts have been made to capture the creature. In July 1964, a fisherman tried unsuccessfully to hook it using whole chickens and lumps of liver as bait. A company called Big Fish Unlimited, offered a reward of $1,400 for the capture of the monster, or any fish over fourteen feet (4 m) long, which might be taken for a monster. As yet no one has come forward to claim the money.

A more recent sighting in August 1983, concerns a different kind of monster. It seems six boaters reported seeing a "giant fish," twenty-five to thirty feet (7.6 to 9 m) long, break the surface of the lake and cross in front of their boat in Yellow Bay. "Its fin was about two feet (61 cm) out of the water and was cutting the water like a shark," Dan Knight told the Associated Press. When it passed the bow it "sent a wave off that would put my fifteen foot (415 m) boat to shame." The six witnesses watched the object for about four minutes.

Flathead Lake is the home of the largest sturgeon ever caught in Montana; in 1955. It weighed 181 pounds (82 kg) and its length was ninety-two inches (2.7 m). The largest sturgeon ever caught in the United States is purported to have weighed 407 pounds (184.6 kg).

Waterton Lake, Montana

Long necks, similar to the Flatheads, supposedly inhabit Waterton Lake, Montana. They have kept boatmen on the Upper Lake in Waterton Lakes National Park, on the hop for many years. Local people call them "Oogles-Boogles." Like those in Flathead Lake, they are credited with lengths of up to sixty feet (18 m). Most descriptions usually concern cow-like, rather than serpent-like heads that sit on the end of long, slender necks with humps behind. Ivan Sanderson researched this plesiosaur type extensively and believed they exist in nearly every country in the world.

Reports from Waterton Lake show considerable evidence in support of "baby monsters," something otherwise only reported from Loch Ness, in 1937. Reports of baby monsters of varying lengths have been pouring in for decades. Details of their external appearance also vary greatly, although the general description remains the same.

Bear Lake, Utah

An ancient family of monsters lives in Bear Lake, spanning the state line between Idaho and northeastern Utah. Twenty miles long (32.2 km) eight miles wide and 200 feet deep (60 m) the sapphire blue lake nearly fills the narrow cleft in the Wasatch Mountains—the mighty western wall of the Rocky Mountains. Here, the land has the feel of distant eons and the look of creation still in progress.

In 1863, the first towns were settled on the west shores of Utah's Bear Lake. The settlers soon became familiar with the legends Shoshone and Bannock told about the lake and the terrible race of monsters who lived in its waters. Long, long ago, the legend said, a young brave had fallen in love with a young woman of an enemy tribe. Both sides were greatly angered and set guards to watch the pair.

The lovers met in secret, but were discovered one day by both tribes, who pursued them along the shores of Bear Lake. Just as they were about to be captured, the Great Spirit changed them into enormous fish-like creatures and sent them into the lake. Henceforth, from time to time the pair and their descendants would take revenge upon the tribes by devouring any unsuspecting victims who happened to venture into the water.

Though the monster had not been seen for many years, since the buffalo inhabited the valley, the Shoshone still viewed the lake with a kind of terror.

That is one of the legends passed on by the Indians. But the whiteman did not need the Indian to tell him there was something strange and uncanny about the lake. Nearly every summer someone saw something unusual in its waters. But since the witnesses were usually alone at the time of the sighting, little attention was paid to reports.

The *Deseret News*, July 27, 1868 published the first newspaper report. The sighting was by a respected local resident. Attitudes changed overnight.

Mr. S. M. Johnson of South Eden, was going along the east shore of the Bear Lake, to the Round Valley settlement. When about halfway, he saw what he at first thought was the body of a drowned person in the lake. He rode to the beach where the waves were running pretty high and waited for the body to be washed ashore. Some minutes later, two or three feet (approximately 1 m) of some kind of animal unknown to him appeared above the surface. It looked like a head and part of a neck. There were ears, or bunches on the side of the head, nearly as large as a pint cup. As the waves dashed over the head, the creature threw water from its nostrils and mouth. It did not drift with the tide, and except for turning its head, remained perfectly still. Mr. Johnson concluded the creature was lying on the bottom of the lake.

A similar monster was seen the next day in the same place by a man and three women. It was described as "very large" and "swimming much faster than a horse could run on land."

Those who had seen the monster before made reports. On the previous Sunday, N. C. Davis and Allen Davis, of St. Charles; Thomas Slight and J. Collings of Paris, Utah; and six women, were returning from Fish Haven, to St. Charles. About halfway they suddenly noticed a peculiar motion or wave in the water some three miles out. The lake was not rough, only a little ruffled by the wind. Mr. Slight said, "he distinctly saw the sides of a very large animal that he supposed to be not less than ninety feet (27.5 m) in length." Mr. Davis didn't think he saw any part of the body, but judging by the size of its wake, he was positive it could not have been less than forty feet (12 m) long.

A few minutes later a second, much smaller animal followed. Mr. Slight likened this second animal in size to a horse.

> A larger one followed this, and so on, till four large ones in all, and six small ones had run southward out of sight.
>
> One of the large ones, before disappearing, made a sudden turn to the west a short distance; then back to its former track. At this turn Mr. Slight says, he could distinctly see it was a brownish color. They could judge somewhat of (the animals') speed by observing known distances on the other side of the lake...all agreed that the velocity with which they propelled themselves through the water was astonishing. (Messrs. Slight and Davis) represent the waves that rolled up in front and each side of (the animals) as being three feet high (1 m) from where they stood.
>
> This is substantially their statement as they told me, (the reporter concluded). Messrs. Davis and Slight are prominent men, well known in this country and all of them are reliable persons whose veracity is undoubted. I have no doubt they would be willing to make affidavits to their statement.

The likeness to the Loch Ness monster—the large wakes, high speed, color and sharp changes in direction attributed to these creatures, is further indication that the same type of animal inhabits both places.

The *Deseret News* carried another report later that year. Marion Thomas and three brothers named Cook saw a monster while fishing opposite Swan Creek:

> Its head resembled that of a serpent and the twenty feet (6 m) of its body which they saw was covered with light brown fur like that of an otter. Two flippers extended upwards from the body which were compared to the fishermen's oars, and they affirmed that they came so near it that they might have shot it with a rifle.

At least three dispatches appeared in Salt Lake newspapers during 1871, giving further characteristics and mysterious appearances of the Bear Lake monster. One dispatch even went so far as to report the capture of a young monster near Fish Haven. The twenty foot (6 m) long creature propelled itself through the water by the action of its tail and legs. It had a mouth big enough to swallow a man. What became of this specimen is a mystery, but there is always the possibility of a hoax.

In 1883, the monster of Bear Lake was immortalized in song by local residents. It remains today as one of Utah's few genuine folk ballads:

Good people, have you heard of late
Of times in Bear Lake Valley?
They're mustering all the forces there
'Tis possible to rally,
And drilling them both night and day,
In spite of Uncle Sam,
To put a fearful monster down,
At first they thought him sham.
The telegraph can't near compare
With his speed of locomotion,
Why, bless my soul, you can't say scat
Before he's cross the ocean.
This fish lays part in Utah,
And part in Idaho,
But which state claims the most of him,
I don't pretend to know.

Great Salt Lake, Utah

Across two mountain ranges to the west, a relative of the Bear Lake monster lives in Great Salt Lake, Utah. This largest natural lake west of the Mississippi, is seventy-five miles (121 km) long by fifty miles (80.5 km) wide. It is the saltiest body of water on earth. Five times saltier than the ocean, Salt Lake is a remnant of the great prehistoric freshwater Lake Bonneville. This vast inland sea, was once 1,000 feet deep. In drying up, it left an enormous deposit of salt and other minerals.

Unlike the Bear Lake inhabitants, the Great Salt Lake monster seems to be more of a land-lubber. About 10:00 P.M. on the night of June 8, the *Salt Lake Semi-Weekly Herald*, July 14, 1877 reports, the night hands at Barnes and Company's salt works at Monument Point near Kelton, suddenly heard strange noises coming from the briny waters of the lake. Then they saw a huge mass of hide and fin rapidly approaching the shore towards the camp.

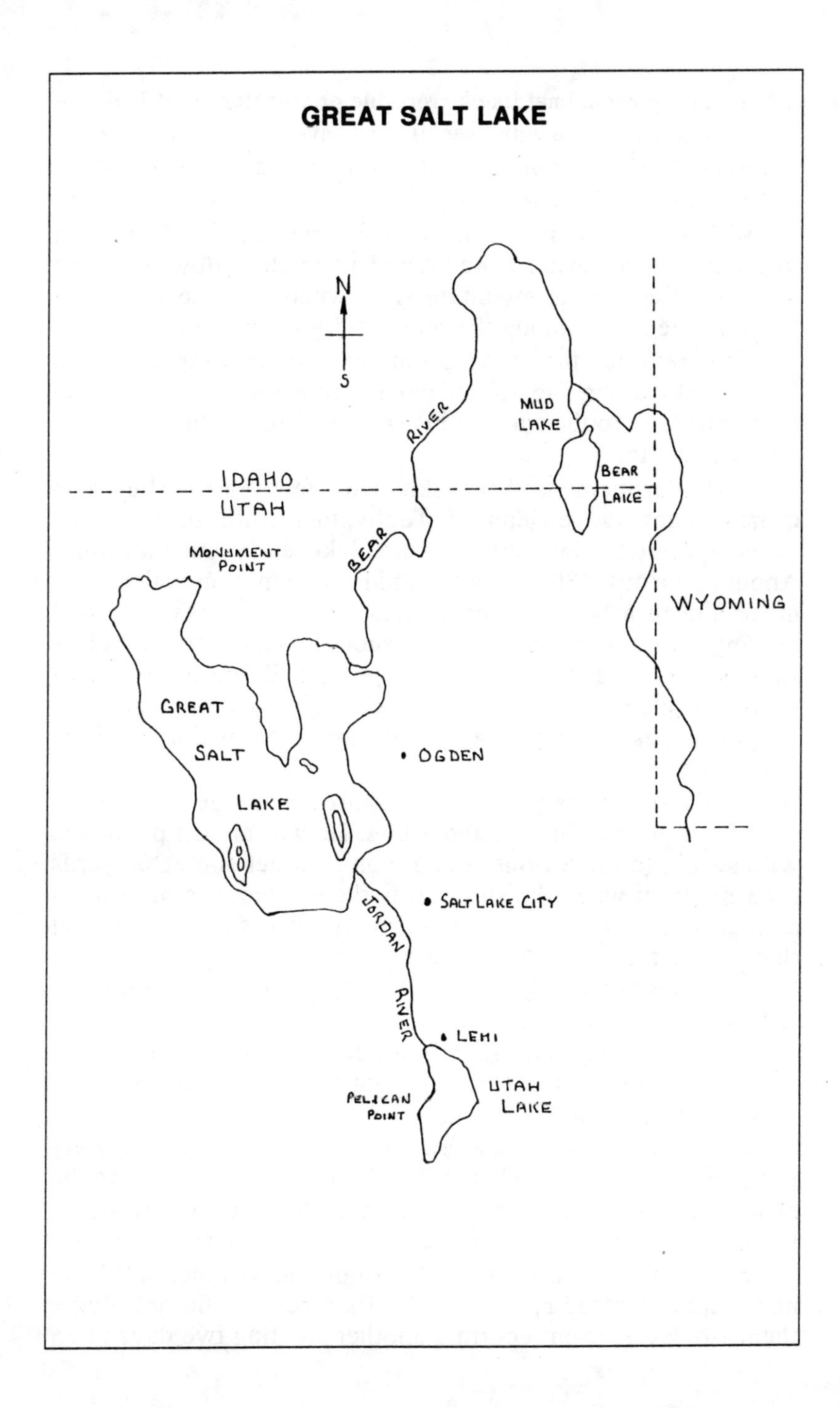
GREAT SALT LAKE
N
S
MUD LAKE
RIVER
BEAR LAKE
IDAHO
UTAH
MONUMENT POINT
BEAR
WYOMING
GREAT
SALT
LAKE
OGDEN
SALT LAKE CITY
JORDAN
RIVER
LEHI
PELICAN POINT
UTAH LAKE

> It was a great animal like a crocodile or alligator...said J. H. McNiel, one of the witnesses. It must have been seventy-five feet (23 m) long; but the head was not like an alligator's it was more like a horse's.

When the creature came within a few yards of shore it raised its enormous head and uttered a terrible bellow. The men promptly fled up the mountain side, where they spent an uncomfortable night among the rocks and greasewood.

They returned the next morning to find the camp upset and huge tracks along the shore, but nothing else. J. H. McNiel made affidavit to the truth of his story before the Justice of the Peace of Kelton.

A similar incident, that took place a few years earlier, also came to light. Judge Dennis J. Toohy, then editor of a Corinne newspaper, was swimming in the lake early one morning. About half a mile (8.04 m) out, a sudden commotion in the water made him turn back fearing he had ventured too far and was nearing the maelstrom; the great subterranean outlet which is supposed to exist somewhere in the lake. Being an excellent swimmer, the judge made good progress, but it appeared to him the water was becoming more and more disturbed as he drew nearer the shore.

About 100 yards (91 m) from shore, the judge stood up in shallow water and looked about him. He was almost paralyzed with shock. In the furious commotion of water some 200 yards (183 m) from where he stood, was the outline of a monstrous animal. Owing to the density and shallowness of the water at that point, the creature's progress was slowed. Its huge tail lashed the water causing waves to rise as though a storm had struck the water.

One glance was enough! The judge made haste for dry land. He grabbed his clothes from the bath-house and dignity aside, set off at a fast run for the city.

He related his story to a few close friends who immediately returned with him to the lake. They found huge tracks along the shore, similar to the ones at Monument Point. Fearing ridicule by the general public, the judge kept his story quiet.

He was right in his assessment of public attitude. Salt Lake newspapers scoffed at J. H. McNiel's report. It did not hinder them, however, from reporting another sighting two days later.

The "huge creature of the reptilian order" was seen in Utah Lake. This lake is also a remnant of Lake Bonneville.

Scientific minds began working overtime. Was it possible these monsters were the same species and all came from Bear Lake? It seemed perfectly possible that such a huge creature could travel from Bear Lake, through the Mud Lake swamps to Bear River, down the Bear to Great Salt Lake and eventually up the tepid Jordan River to Utah Lake. There was no occasion for further study as the monsters promptly went into seclusion and were not heard from again for five years.

The year 1882 sparked off a new series of sightings at Utah Lake. "It has frightened men—and, far better evidence than that," wrote one travel author, "it has been seen by children when playing on the shore. I say 'better', because children are not likely to invent a plausible horror in order to explain their sudden rushing away from a given spot with terrified countenances and a consistent narrative."

But by 1883, much to the satisfaction of Bear Lake residents, their monster was back where he belonged, in Bear Lake. They didn't take kindly to one of their monsters going on a jaunt across the country and bestowing fame upon other lakes.

Theories suggested the monster had followed an established migratory route between Bear and Utah Lakes. However, if such a migration pattern did exist, it was severely disrupted in the 1920s, when four dams were built in Bear River. Between Bear Lake and Great Salt Lake, the dams diverted much of the water for irrigation, culinary, and industrial use. This lowered the level of Great Salt Lake to the extent that the lower courses of both the Bear and Jordan Rivers almost dried up.

That is the modern explanation for the absence of sightings in Great Salt Lake and Utah Lake. Monsters in the Bear Lake still manage to keep their enthusiasts happy, with continued sightings into the 1990s.

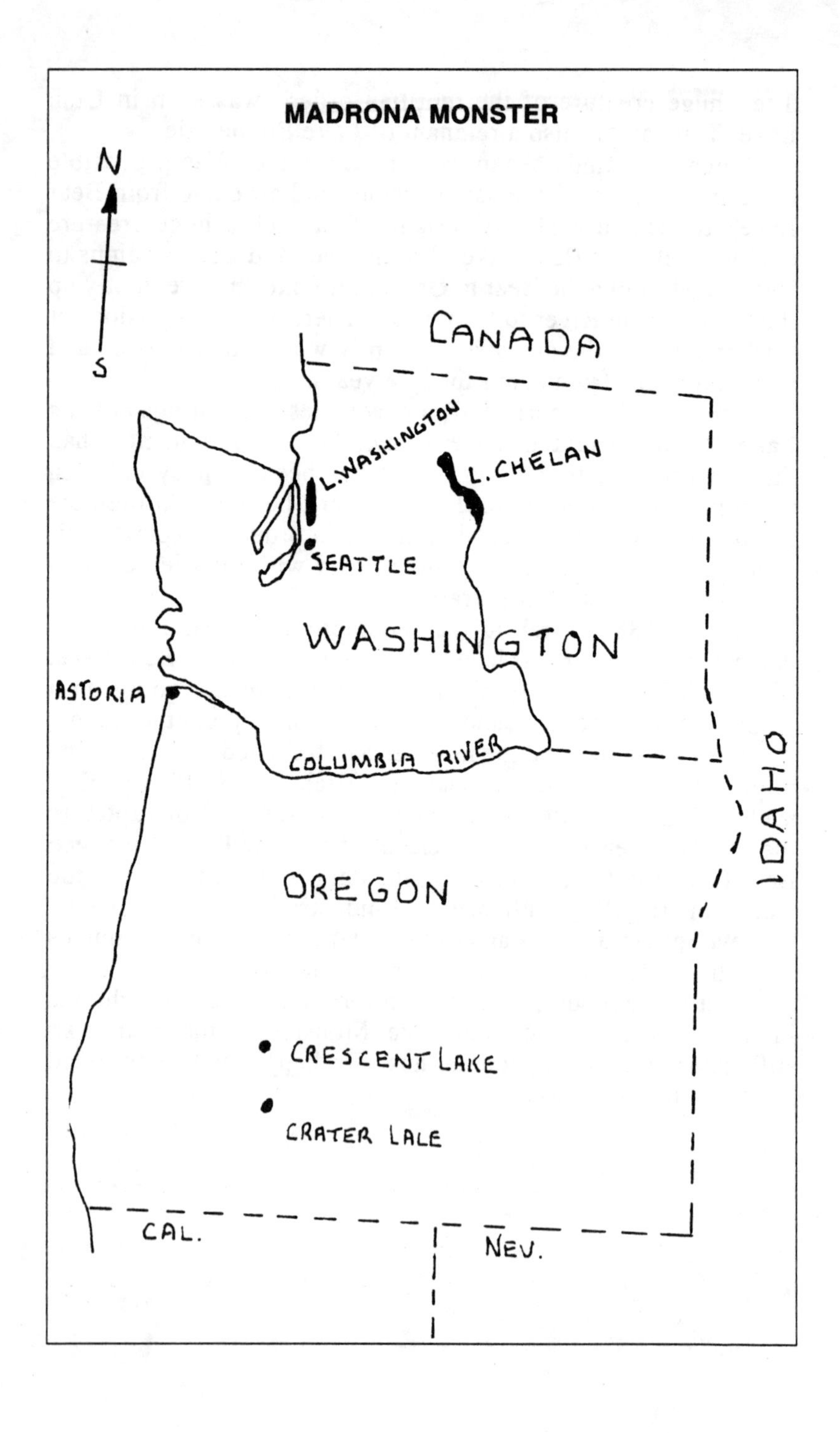
MADRONA MONSTER
N
S
CANADA
L.WASHINGTON
L.CHELAN
SEATTLE
WASHINGTON
ASTORIA
COLUMBIA RIVER
IDAHO
OREGON
CRESCENT LAKE
CRATER LALE
CAL.
NEV.

10

The Madrona Monster

Lake And River Monsters Pacific Coast States: Washington; Oregon; Alaska

The Pacific Coast states region is famous in all aspects of the North American monster scene. Not only is it prime Bigfoot territory and the purported home of the Flying Phantom, it also has its share of water monsters.

Lake Washington

Washington, the northernmost state in the region, is host to the Madrona monster. This creature resides in Seattle's Lake Washington; once again, a large glacial lake formed when glaciers scooped out the land and water filled the hollow places.

The Madrona monster is described as seventy-five to one hundred feet (23 to 30 m) long and is usually seen on hot sunny days in July. But in February, 1947 he made an early appearance. The *Daily Gazette* reported the monster had appeared for the "third time" at the south end of the lake, as "a sudden disturbance in the water followed by the violent surfacing of a tailless object." Another witness saw "a dark, crinkly backed object moving south in the lake."

Mrs. Mary Barrie, a Madrona Beach housewife, saw the

monster on two separate occasions romping in the water near her lake front home. Her gardener also saw it. "It was either a monster or a submarine..." Mrs. Barrie told the *Seattle Times*. Naval Reserve officials said that no submarines were maneuvering in the lake.

About 4:30 P.M., April 6, 1964, a retired Army Colonel, Henry B. Joseph, his wife, six-year-old son and cocker spaniel "Suzie," were rounding the north end of Mercer Island in their boat when the colonel noticed something big, black and shiny about 300 yards (274 m) off their port bow. At first he thought it was a thirty foot (9 m) log, but closer inspection revealed otherwise. "In my thirty years of service, I've seen sharks, whales, blackfish, porpoises, and manta rays, but nothing like this," he said. "Suzie was barking like mad. She seemed to go nuts!"

After passing the monster, the colonel circled around and came back for a second look. But the animal had disappeared. Some forty-five minutes later they saw "an end of it" sticking up out of the water. "We kept our eyes on it and headed straight for it at ten knots," he said. "When we were within 100 yards (91 m) or so, the creature just disappeared beneath the surface."

Crater and Crescent Lakes, Oregon

Crater Lake, in the Cascade Range in southwest Oregon, is one of the great scenic wonders of the world. It was formed about 7,000 years ago, after the glacial period, when the top of a prehistoric volcano, Mount Mazama, collapsed and was swallowed up inside the mountain. Deep blue in color and circular in shape, 1,983 feet (605 m) deep, the lake has no known outlets and no streams flowing into it. Rain and snowfall keep its level almost constant. Crater Lake National Park, established in 1922, surrounds the lake.

As one would expect of such a mysterious place, there are many legends attached to the lake. Early Klamath Indians believe Crater Lake was a haunt of water-demons who dragged into it and drowned all who ventured near. They saw the lake as the work of the Great Spirit and all but one man who looked into its circular basin and sheer walls, stole away. But the one man felt that if the lake was the home of gods they might have some message for mankind. He camped on the brink of the lofty cliffs and waited.

In his sleep he had a vision and heard voices, but could not understand their meanings. The same dream occurred every night, till finally he went down to the water and bathed. Instantly his strength increased and he saw the people of his dreams in the waters. Whether they were good or bad, he did not know.

One day he caught fish from the lake as food. Instantly a thousand water-devils emerged from the lake and seized him. They carried him to a rock, on the north side of the lake, that rose 2,000 feet above the water and dashed him to his death. Then they gathered his shattered remains from below and devoured them. Having once tasted men's blood, the legend says, the genii of the waters have been eager for more.

Another rock, on the south side of the lake, called the *Phantom Ship*, is believed by the Indians to be a destructive monster. So much for legends! But there have been sightings of strange creatures in both Crater and Crescent Lakes since white people came to the area. Crescent Lake lies thirty miles (49 km) north of Crater Lake.

Alaska

Alaska, the northernmost state in the union, is not without its share of monsters.

Iliamna Lake

Several monsters are purported to live in Iliamna Lake, west of Cook Inlet, at the base of the Alaska Peninsula, near the Iliamna Volcano. Seventy-five miles (121 km) long and ten to twenty-five miles (16 to 40 km) wide and in places over 1,300 feet (396 m) deep, the lake has an area of 1,000 square miles (2,590 sq. km). It drains into Bristol Bay by Kvichak and is noted for rainbow trout. The lake has a reputation for being very rough and very dangerous. With mountains north and south, winds can thunder across the lake and set up a storm on the surface that no boat could survive. The waves roll like an angry sea and no one will venture out on the lake.

The Kenai Indians along the barren lake maintain they have seen a monster so huge, so frightening and so powerful that the famed Loch Ness monster pales by comparison. They believe

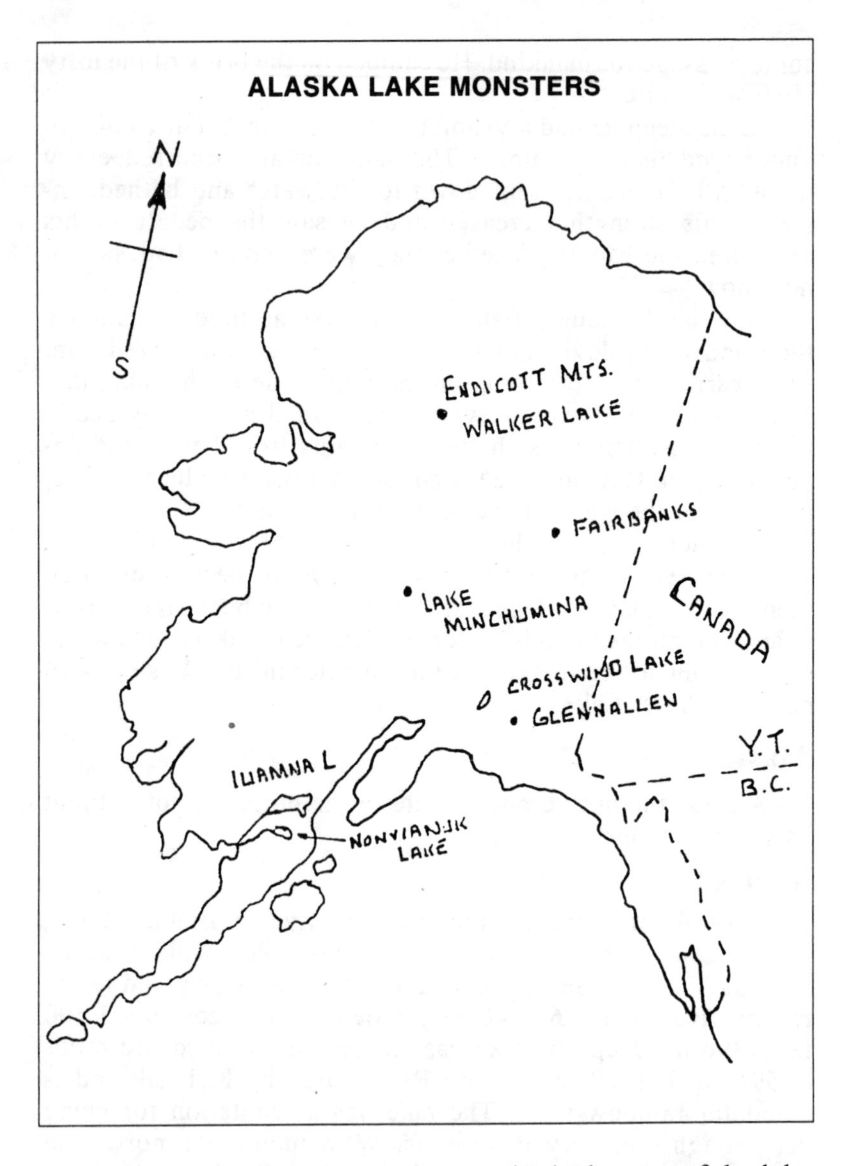

the monster has plowed great furrows in the bottom of the lake making it more than 1,000 feet (305 m) deep. Almost all of the monster's bad moods come after sunset. During the day he is noticeably quieter. When the lake is calm, the Indians say he is sleeping.

The Aleut Indians have been familiar with the monsters of

Iliamna Lake for a long time and view them with considerable respect. Earlier in the century, a monster upended one of their boats and swallowed a crewman. For a long time afterwards the Indians carefully avoided that part of the lake. To this day many of the natives will not venture onto the lake in a boat with a red hull. They believe the monster is angered by this color and will attack.

No scientist has ever gone to the lake to investigate reports, but adding a measure of credibility to the Indians' stories, pilots have told of seeing huge mysterious shadows beneath the surface as they flew over the lake. The dull aluminum colored shadows mysteriously disappeared when the pilots swooped lower to have a closer look.

Other witnesses have described the monsters as approximately twenty feet (6 m) long and possessing broad blunt heads, tapered bodies and strong tails. As is the case with most deep lakes, Iliamna Lake was once part of the sea. Fishermen believe the monsters are some kind of prehistoric fish—probably of the shark family. They were stranded in the lake centuries ago when the land closed around the lake and adapted to the conditions therein.

Most sightings occur during September and October. No carcass has ever been washed ashore. But this is not surprising. In the extremely cold water of the lake a carcass would probably sink quickly to the bottom.

There is a $1,000 bounty on the Iliamna monster's head. As a result, many attempts have been made to capture him. Using one-sixteenth inch (1.6 mm) stainless steel wire for line, a quarter inch (6.4 mm) thick iron rod bent into a hook and baited with a hunk of moosemeat and a fifty-five gallon (250 l) oil drum as bobber, the hopeful fishermen have waited patiently. But each time the monster strikes he takes not only the meat but all the tackle as well!

Fish and game authorities suggest the monsters are actually land locked sturgeon. Before they were fished out from northern rivers and the Great Lakes, sturgeon were taken measuring as much as twenty feet (7.6 m) long and weighing 2,600 pounds (1,180 kg).

Whatever the identity of Iliamna Lake's mysterious inhabitants—giant sturgeon, lake trout, beluga whales, sharks or something left over from prehistoric times, something capable of snapping a stainless steel cable as though it were thread has to be a monster!

Similar monsters to the Iliamna ones are also said to inhabit Crosswind Lake, near Glen Allen; Lake Minchumina between Fairbanks and Mount McKinley; Walker Lake east of Kotzebue and Nonvianuk Lake, southwest of Homer.

Ogopogo

11
Monstrous Canadian Cousins

Lake And River Monsters

Canada, particularly the north, is dotted with thousands of lakes, some very deep and remote, where monsters could exist virtually unnoticed and undisturbed. Across the southern more populated regions, reported sightings echo from coast to coast. Some of these strange phenomena have come to rival their famous contemporaries in Loch Ness and Lake Champlain.

Nova Scotia

Lake Ainslie

Many reports of serpent-like monsters have come from Lake Ainslie, near the west coast of Cape Breton Island. Dr. Carl Medcof, a retired research scientist formerly with the St. Andrews Fisheries Biological Station, suggested the monsters are really balled-up eels—hundreds of them in one mass, like worms dumped out of an angler's tin can. Dr. Medcof observed them himself. He said eel-balls are apparently seen in deep water in late summer. They can measure as much as six feet (2

m) in diameter. They bob up and down in the water in a slow rolling (undulating) motion. It would not be difficult to mistake them for sea serpents.

Similar serpentine monsters have also been reported from the salt water, Bras d'Or lakes.

New Brunswick

Dungarvon River, a tributary of the Miramichi River

New Brunswick claims a perfectly good, if somewhat atypical, monster in the "Dungarvon Whooper." Its occasional howls in the Miramichi woods and along the banks of the Miramichi River "have sent chills up the spines of lumberjacks for more than a century." Naturalists are inclined to think it is the cry of an eastern panther which pierces the air like a woman's scream. But guides who know the panther's shrill call think otherwise!

The Monster of Lake Utopia

Lake Utopia, Charlotte County

Seven miles long, four miles wide and in some places 400 feet deep, Lake Utopia is the home of a venerable monster named "Old Ned."

The first reports were recorded in 1872 and concerned a monster resembling a "huge black rock," but it moved and churned all the time. At least ten feet (3 m) wide and approximately sixty feet (18 m) long, Old Ned has also been likened to a huge sea-turtle or dragon.

People living by the lake, nearly without exception firmly believe that a huge fish or serpent has a home in the lake. They have seen it basking full length, like a huge pine log on the surface of the lake. Sightings usually occur after the winter ice breaks up.

Lake Kilarney

Another unconventional lake monster, known as the Coleman Frog, made its "happy hopping ground" in Kilarney Lake in the early 1880s. Apparently, the creature was five feet four inches (1.5 m) long and weighed forty-two pounds (19 kg).

More than a century ago, Fred B. Coleman, proprietor of Baker House, a Fredericton hostelry, would amuse guests with stories about the monstrous bull-frog. Guests coaxed the creature out of the lake with June bugs and even Scotch Whiskey. But Coleman attributed its remarkable size to its consumption of huge portions of locally churned buttermilk.

The giant amphibian thrived until 1885, when fish poachers exploded dynamite in the lake. In the process the frog was killed. Coleman retrieved the body and had it stuffed. He displayed it in the lobby of the Barker House. In 1959, his family donated it to the York Sunbury Historical Society Museum in Fredericton, where it currently resides in a glass case.

The Coleman frog gained international fame when it was drawn and described by Robert L. Ripley in his well-known feature "*Believe It or Not*!"

Quebec

Black Lake

The unknown inhabitant of Black Lake is described as a gigantic serpent twenty to thirty feet (6 to 9 m) long, with a body the size of a stove-pipe. There were reports of sightings in 1894 and 1896 by several witnesses including fishermen and a trapper. They stated, it "lashed the water and crossed the lake above the island...making apparently, for the White Stream where it was lost to view..."

Lac Memphremagog

"Memphre" is the name of the snake-like monster in Lac Memphremagog, in Quebec's eastern township. He was reportedly seen by 137 people on seventy-one occasions between

1816 and 1987. Described as being more friendly than monstrous, Memphre put in an appearance at 5:00 P.M. on August 12, 1983. Barbara Malloy and three other witnesses were in a car at the top of the hill overlooking Lac Memphremagog, near Horseneck Island. They had stopped to look at the scenery. In the water, about a mile (2 km) away, they saw a creature swimming extremely fast. In its wake it left waves "like a speed boat would make."

According to Mrs. Malloy, the creature was dark brown and had a head like a horse's, with a long neck. She estimated its size to be "longer than a house." She and the others watched for two minutes until the animal swam out to the middle of the lake and disappeared.

The United States/Canada border slices through the lower end of Lake Memphremagog. But Memphre apparently considers this no concern of his, and swims freely between both territories.

The real expert on the Lake Memphremagog monster's activities is Jacques Boisvert of Magog, Quebec. Boisvert cofounded, in 1986, with Mrs. Barbara Malloy of Newport, Vermont, the International Society of Dracontology. Boisvert's research indicates sightings of a monster in the lake date back to Ralph Merry, the first pioneer to settle in Magog (Outlet), in 1798.

The monster has a long history. There were at least seven sightings in 1994, involving twenty-one persons.

Mocking Lake

Reports of a monster in Mocking Lake reach back for almost 100 years. It is described as an unknown animal twelve to eighteen feet (4 to 6 m) long and brown or black in color. It has a round back two or three feet (1 m) wide and a sawtooth fin down the center.

Lake Pohenegamook

Lake Pohenegamook lies 280 miles (450 km) northeast of Quebec City, near the New Brunswick border. Over the past fifty years "Ponic" has been reported by dozens of eye witnesses, including a local priest and a former mayor. The latter said the creature looks exactly like drawings of the Loch Ness Monster.

"I was standing near the lake when I saw it loom out of the water. It was very dark gray and at least forty feet (12 m) long. It stayed afloat for a minute, then dove straight down, leaving a trail of waves."

Ponick was the town mascot during its 1974 centennial celebrations. His likeness, drawn by a local artist, graced the official ceremonial handbook. Sightings date back to 1874.

"Ponic," the gigantic creature sighted in Lake Pohenegamook.

Ontario

Berens Lake

Trappers and Indians have reported sightings of a strange denizen resembling an alligator in this lake. It purportedly lives in the water, emerges occasionally to slither around on land, then slides back into the lake again. Some trappers are afraid to approach the lake and have asked the Provincial Government to organize an expedition to hunt down and destroy the creature.

Mazinaw Lake

Mazinaw Lake is thirty miles (48 km) southeast of Bancroft. In its 300 foot (91.4 m) depths, there lives a formidable monster. It has claimed one life and has forced hundreds of bathers to flee in panic. Biologists believe the creature might be a huge sturgeon—ugly creatures which can live to be hundreds of years old and weigh up to 3,000 pounds (1,362 kg).

Muskrat Lake

This lake lies sixty miles (96.5 km) west of Ottawa. It is ten miles (16 km) long and in some places as deep as 240 feet (73 m). Samuel de Champlain wrote about its watery inhabitant in the 1600s. More recent reports in 1968, 1971, 1976 and 1988, suggest "a large aquatic animal" perhaps similar to the Loch Ness monster, about twenty-four feet (7 m) long, silver-green or cocoa-brown in color, with two humps and a green fin down its back.

The monster is variously known as "Mussie," or "Hapyxelor"—because of its friendly nature! Two 1988 sonar sightings showed two eight to ten foot (2 to 3 m) objects swimming side by side.

Nith River

Members of the small farming community of New Hamburg, located between Kitchener and Stratford, complained to the chief of police about a strange creature which was described as large and lizard-like. When it was seen by the chief of police himself, he vowed that he would either capture it or, if necessary, shoot it. The town clerk, treasurer and others described it as green-brown in color, with a scaled tail, about fifty pounds (22.6 kg), four-legged, and three-toed.

Lake Ontario

In 1835, a German science journal published an account of a serpent in Lake Ontario. Kingston, at the junction of the Great Lakes and the St. Lawrence River, has enjoyed more than its share of sightings, with reports in 1829, 1867, 1881, 1888, 1892 and 1931. The monstrous inhabitant, twenty-five to forty feet (8 to 12 m) in length, was nicknamed "Kingstie" in the 1960s.

Of the 1881 report, the *Kingston Whig*, September 14 that year, stated the monster was seen making its way into the Rideau Canal and surprising the crew and passengers of the steamer *Gypsy*, en route from Ottawa to Kingston: "The sportive creature made its appearance, of course, unexpectedly, and annoyed the people with its immense proportions, unsightliness and graceless pranks in the water."

Two doctors, R. R. MacGregor and Frank Bermingham, who saw the beast in 1931, suggest it has chameleon-like qualities, changing to blend with its surroundings. Its skin may be covered with "warts or bunches." It has a long powerful tail and eyes that are fierce and glittering. Despite Kingstie's monstrous size and the *Whig's* derogatory statements, the monster is not considered menacing, but is usually described as playful and endearing.

Kingstie is not the only monstrous inhabitant of Lake Ontario. A summer 1978 Toronto sighting described a long, serpent-like object exceeding fifty feet (15 m) in length, with a scaly body and numerous humps along its back. A similar creature was seen in August, near Oshawa harbor. This creature has affectionately been dubbed "Oshawa Oscar." His statistics are as follows: "about twenty-five feet (8 m) long, with three humps, and gray-black in color."

Yet another report, in August 1978, told of a monster (possibly the same one) in Beaverton Harbor; local residents claimed it for their own and officially christened it "Beaverton Bessie."

Lake Simcoe

There were persistent reports of a monstrous presence in Lake Simcoe, forty miles north of Toronto, by settlers in the nineteenth century. These reports continued spasmodically, but none were taken seriously until July 1963, when a report of a sighting at Keswick, Ontario appeared in the Oakville *Journal Record*.

The creature is described as approximately seventy feet (21 m) long, dog-faced and with a neck the size of a stove-pipe. Sometime in the 1950s the creature was named "Igopogo," after Ogopogo, in Okanagan Lake, B.C., and "Manipogo," in Lake Manitoba.

Another monstrous serpent is said to occupy Kempenfelt Bay, the westward extension of Lake Simcoe. There have been frequent reports since 1881. Somewhere along the way the creature was named "Kempenfelt Kelly"—perhaps because the alliteration makes for a memorable name.

Kelly is generally described as a charcoal gray, serpentine creature thirty to seventy feet (9 to 21 m) long, with several dorsal fins down its back and a fish-like tail. His head is like a horse's or dog's, with a wide gaping mouth and prominent eyes. He is said to enjoy sunbathing.

In 1976, the Barrie-based cartoonist John Beaulieu sketched Kelly as a huge, worm-like serpent with a couple of humps, a button nose and kindly eyes. Shortly thereafter, the Greater Barrie Chamber of Commerce began to promote the monster as a tourist attraction. Now there are even Kempenfelt Kelly t-shirts.

But Kelly himself is a shy monster and has never allowed his photograph to be taken. However, he was caught in a radar sounding by William W. Skrypetz from the Government Dock and Marina (Lefroy), on June 13, 1983, at approximately 3:30 P.M. The Skrypetz sounding shows him to be a long-necked, heavy-bodied creature not unsimilar to Scotland's Loch Ness monster.

There are also indications of other mysterious inhabitants in Lake Simcoe's Kempenfelt Bay and Cook's Bay. Each year visitors and citizens of Barrie, the city which overlooks the western tip of Kempenfelt Bay, report sightings of large, unidentified aquatic creatures in these waters. In the early 1800s fur traders and British soldiers made the occasional sighting. But it was not until World War I that widespread reports accumulated.

Perhaps Igopogo and Kelly are just two of a rising family of monsters that make Lake Simcoe their home.

Manitoba

Lake Manitoba

Manipogo is the name given to the monster (second in the 'Pogo group) of Lake Manitoba, Lake Winnipegosis and Dauphin Lake. Sightings recorded since the beginning of the century describe a creature forty feet (12 m) long "with diamond-shaped" head about eight inches (20 cm) wide...and a single horn "like a periscope." A "large hump and a smaller one at the end" have also been noted. The color is generally given as brownish black and glistening. It moves at approximately fifteen miles (24 km) per hour, leaving a wake of about eight feet (2.5 m) and fifteen inches (38 cm) high, behind the head.

On August 12, 1962, an American television commentator managed to photograph the monster. The picture is generally considered genuine.

Another widely reported sighting was made in August 1964, by Ralph Sanderson and his family, north of Toutes Aides. They spotted a two-humped, greenish black creature with a tail. The creature was perhaps sixteen feet (5 m) in length. It surfaced and outswam their outboard motorboat before disappearing in the distance. Although there are earlier and more recent reports, the sightings of the sixties were richer and more plentiful.

Saskatchewan

Rowan's Ravine

A long, dark and shiny serpent was reported from Rowan's Ravine, fifty miles (80 km) northeast of Regina, in July 1964. There were several witnesses at the Provincial Park who said the creature was about thirty feet (9 m) long and "looked like egg-shaped groups attached together," as it swam through the water.

Turtle Lake

Indian legends tell of "the big fish" that lives in this picturesque resort lake, fifty miles (80 km) east of Lloydminster. It is described as being about twenty-five feet (8 m) long, with a fin and three bumps on its back, a long neck and head like a horse, pig or dog. Stories have been told about this monster for over sixty-five years.

In winter the "Turtle Lake Terror" makes itself known by tearing huge holes in the nets of local fishermen. In summer it frightens swimmers and boaters. The creature has been dismissed as being a large sturgeon, among other things. But, perhaps it is a remnant of the past—an aquatic plesiosaur that lived in Saskatchewan millions of years ago, when the province was covered by sea.

Alberta

Battle River

Battle River, seventy-seven miles (124 km) southeast of Edmonton, is said to house a serpent twenty-five to thirty feet (7 to 9 m) long. It is described as "gray in color with a large head and neck tapering from eight or nine inches (22 cm) at the head, to a body the size of a stove-pipe. Near the center there was a big bulge as though it had eaten a big meal." The creature swims at approximately ten to fifteen miles (16 to 24 km) per hour, lashing the water to a foam.

Saddle Lake

About 100 residents have reported sightings of a monster in

Saddle Lake, 100 miles northeast of Edmonton, over the past decade. It is generally described as serpent-like, eight to eighteen feet (2 to 5 m) long, with horse-like head and hairy torso.

British Columbia

Harrison Lake

Harrison Lake lies eighty miles (129 km) east of Vancouver. It has an average depth anywhere from 800 to 1,000 feet (244 to 305 m) and in some places more than 3,500 feet (1,067 m); some people say it is bottomless! In fact, it is just the place for a monster. Stories abound among Indians and non-Indians alike. But a clear description of the monstrous inhabitant is hard to come by at this time.

Lake Kathlyn

The Indians have an interesting legend about Lake Kathlyn, formerly Chicken Lake, in the Bulckley Valley. Long ago, the legend says, an Indian princess was seized by a serpent "larger than a canoe," while she was swimming in the lake. The Indians surrounded the lake with boulders and logs in an effort to save their princess. They set fires under the boulders and pushed them into the water, thereby boiling the monster alive. The body of the dead princess was recovered and three days later the lifeless monster rose to the surface of the lake.

So much for legend, but there was a report of an unidentified serpent in Lake Kathlyn, in 1934.

Okanagan Lake

Okanagan Lake, on the Pacific slopes of the Rockies is famous as the home of Ogopogo, first in the 'Pogo group of monsters. Stories have been told about this monster since before the first missionaries set foot in the valley, in 1860. The Indians called him "Naitaka." They feared the monster as a creature from an evil world and felt it necessary to appease him with gifts of a chicken or pup before venturing out on the lake.

Early white settlers also became familiar with the monster. At first they kept their experiences to themselves, fearing ridicule, but as time moved on and the Okanagan Valley became more populated, attitudes changed. The monster was renamed Ogopogo, as reports became more frequent.

Today, this dark monster of the deep is respected by both

Indians and non-Indians, with repeated sightings since the beginning of this century. Descriptions usually concern a series of humps moving swiftly, caterpillar-fashion, through the lake, leaving a heavy wake. Other information gathered over the years suggests a serpent-like creature forty to sixty feet (12 to 18 m) long, with sleek black or dark green body and a head like a horse or sheep. He also sports whiskers, a forked tail and possibly flippers for locomotion.

Sightings of British Columbia's famous monster always occur in the warm summer months. It is believed he hibernates during the winter months, October to May, in a cave somewhere beneath Monster's Island—a small barren and rocky island at a bend in the lake some twenty-two miles (35 km) up from Penticton.

Ogopogo sightings hit a record high during the summer of 1990. Apart from an aerial sighting in March, all reported sightings took place in the peak viewing months of June, July and August. The first sighting was in early June, when Rutland, British Columbia teacher Natalie Butcher, out boating with about a dozen friends, reported seeing the creature in the waters at Bear Creek Park, north west of Kelowna. A few days later a double sighting occurred—first by Ann Klein at her home near Kelowna's Gyro Beach, at 10:45 A.M. and half an hour later, by Bob Pearson and Ron Grigg. The two men were traveling east on the Okanagan Lake floating bridge when they spotted a thirty foot (9 m) long object swimming diagonally north from the causeway.

In July, a television crew from Japan's Nippon network came to Kelowna to film a documentary on the Ogopogo legend. On their first day (July 24), boat owner Mike Guzzi noticed a peculiar signal on his sonar, again near Bear Creek. He turned the boat around to pass over the area again. Cameraman Steve Sidaway caught the apparition on videotape. It was about thirty-three feet (10 m) long, 328 feet (100 m) below the surface and surrounded by turbulence, like bubbles from a propeller.

A sculpture of Ogopogo resides in Kelowna's lakeside park. Photographs of his monstership have also been taken and his name is engraved in many tourist and commercial business enterprises.

Lake Tagai

Lake Tagai, Prince George, is approximately thirty-five to forty feet (11 to 12 m) deep, except for a 150 foot (46 m) deep hole, just off a point in the horseshoe-shaped lake. There have been rumors and stories told by fishermen for many years about an unknown creature dwelling in those depths.

But, once again, the creature is mysterious. All we know is, it is about ten feet (3 m) long, black and moves with great speed creating large waves in the usually calm lake.

The history of North America's lake monsters stretches across the continent for more than a century and a half. Reported sightings now number literally in the thousands. Witnesses include prominent citizens from many walks of life and of unquestionable veracity.

No doubt some of the sightings could be attributed to mistaken identity—known creatures that, because of their surprise appearance, were not immediately recognized. Others could be attributed to logs, tree trunks, branches, large fish and overturned boats floating in the water. But others clearly cannot be explained away.

Many reports have been from clear, deep water lakes or rivers entering or exiting these. The majority are from lakes gouged out or left by the recent (in geological terms) advances and retreats of ice-caps. Indeed, it is conceivable lake monsters could exist wherever the water is deep and quiet.

In the main, lake monsters resemble the prehistoric plesiosaur, although those in the Great Lakes seem more like gigantic eels or sea-snakes. Given the history of eels, this does not seem improbable.

If sightings of lake monsters seem less in recent years, this is not surprising. Monsters are offended by civilization, its noisy pastimes and activities on the waters. They prefer to gambol where beginnings are made, where the land is still wonderfully new and still open for discovery. Mysterious Idaho, where many hundreds of remote lakes still remain undiscovered, might be the proving ground for North America's lake monsters. In the meantime, "monster buffs" must keep their eyes open and cameras at the ready!

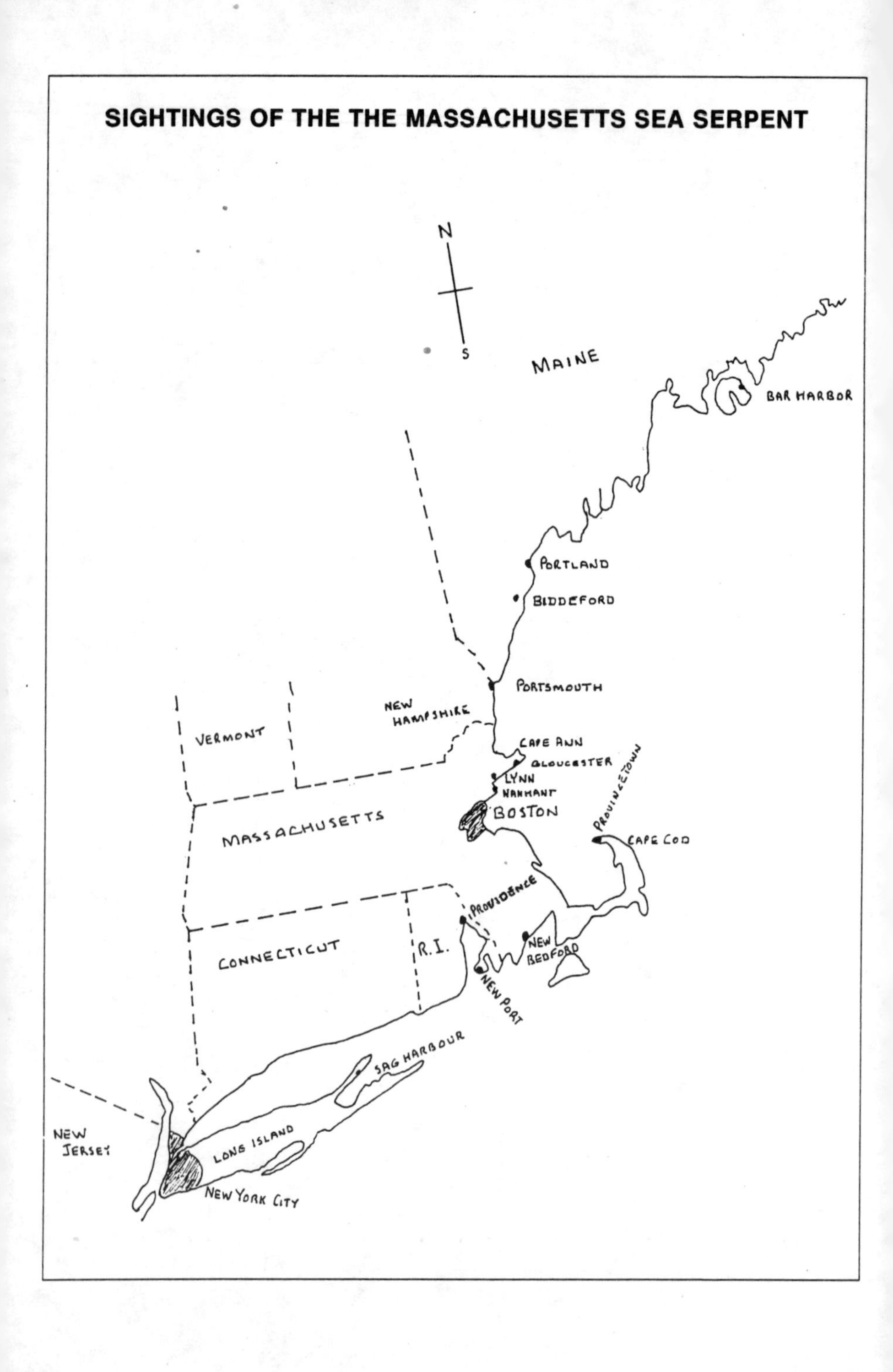
SIGHTINGS OF THE THE MASSACHUSETTS SEA SERPENT
N
S
MAINE
BAR HARBOR
PORTLAND
BIDDEFORD
PORTSMOUTH
NEW HAMPSHIRE
VERMONT
CAPE ANN
GLOUCESTER
LYNN
NAHANT
BOSTON
PROVINCETOWN
CAPE COD
MASSACHUSETTS
PROVIDENCE
CONNECTICUT
R.I.
NEW BEDFORD
NEW PORT
SAG HARBOUR
LONG ISLAND
NEW JERSEY
NEW YORK CITY

12

Sea Serpents—East Coast

The concept of huge, unknown sea beasts living in the oceans of the world is not new. Man has been telling stories about these strange phenomena since the beginning of civilization. Many are the spine-chilling tales of gigantic serpents that rose from the depths to terrify and often engulf great ships.

Ancient map-makers depicted seas teeming with beasts of every description—spiny, winged, horned, sharp-toothed and many tentacled. While most of these can be dismissed as imaginary creatures, a few like the ancient Leviathan (the blue whale) and the Kraken (the giant squid) have made the transition from mythical fancy to fact.

This is not the case, however, with another huge saurian—generally called the sea serpent, which is most often depicted in sea monster stories. Whether or not an actual reptile, the creature is usually described as a giant snake with a neck rising from a slim body with flippers.

There have now been thousands of sightings of this strange phenomenon in every corner of the globe. Witnesses have included sea captains, bishops, judges, priests and other people of responsible position. Many witnesses have signed sworn statements as to the truth of their reports.

Yet, despite this fact and the large volume of information now collected, no carcass or other tangible remains positively identified as a sea serpent, has ever been found. Therefore, science cannot accept their existence.

From time to time, carcasses have been washed ashore and

for a while it has seemed the search for the elusive sea serpent might be over. But then all hopes have been dashed when the remains were identified as those of a basking shark, oarfish or some other masquerader. The basking shark particularly, when in a state of decomposition, looks very much like a sea serpent. The only bony structure this shark possesses is its back bone, which runs from the base of its small skull to the tip of its tail. As it decomposes the flesh and soft cartilage breaks away leaving only a skull, backbone and a little flesh adhering to the fins and tail. The long, bare expanse of backbone between the small head and the tail looks amazingly like the long, thin neck of a sea serpent. The fact that the sea serpent still languishes in the gray area between fact and fiction has not prevented a few free-thinking scientists from writing about it. Over the past 100 years eminent scientists such as Samuel Rafinesque-Smaltz, Anton Cornelis Oudemans, Ivan Sanderson and Bernard Heuvelmans have all contributed important works on the subject.

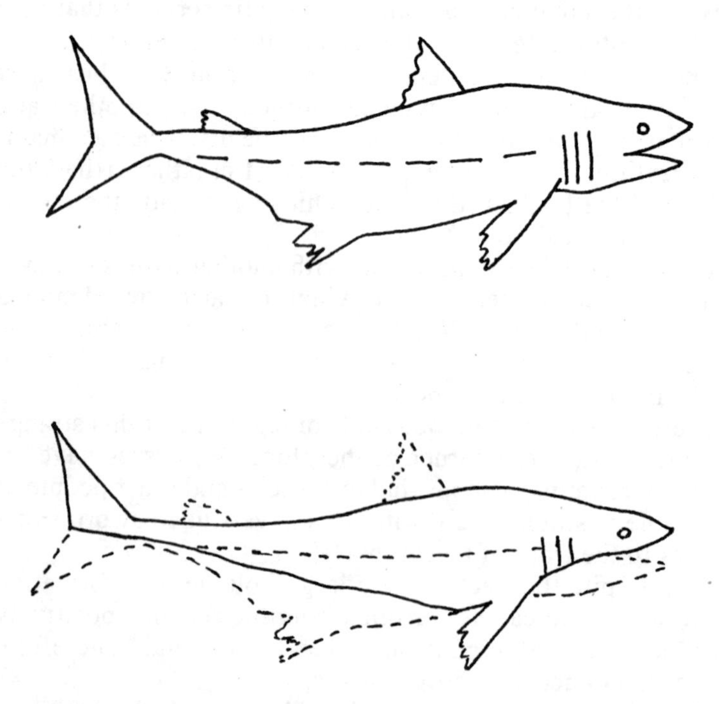

Basking shark—dotted lines show parts that break away in decomposition.

Dr. Heuvelmans' book, *In the Wake of the Sea-Serpents*, published in 1968, was the result of ten years' painstaking research by the author in selecting and crossmatching evidence of literally hundreds of sightings from all over the world. Through the pages of his book we see the sea serpent as a "shy, often inquisitive creature, gamboling in the oceans of the world, living its life as any other, or 'known' creature of the deep."

Long-necked sea serpent

Merhorse

Heuvelmans defines nine different types of sea serpents as inhabiting the oceans of the world; seven of these appear in North American waters: long-necked, a tail-less seal-like animal with a very long neck and fatty humps on its back; merhorse, an animal similar to that above, except that it has large eyes, whiskers and a mane; many-humped, a long mammal with a blunt head, short neck, one pair of flippers, a series of humps along its back and which moves with an undulating movement; super-eel, limbless serpent-like animal, probably a fish; marine saurian, a primitive sea-going crocodile; many-finned, a long jointed mammal with rows of triangular fins and the super-ot-

Many-humped sea serpent

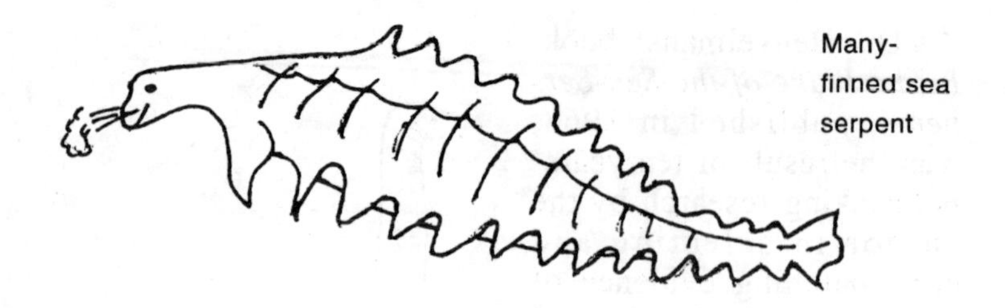

Many-finned sea serpent

ter, a long-tailed mammal resembling a giant otter.

Heuvelmans also tells us sea serpents "are common to most seas." They like fine sunny weather and calm seas. They prefer to swim in warm ocean currents, spending the summer in the Northern Hemisphere and migrating in the winter to the Southern Hemisphere. They vary greatly in size up to 200 feet (61 m) long.

Early American reports came mainly from the area between Cape Hatteras and Newfoundland. There were 130 sightings before 1900, mostly in Penobscot Bay, Maine and Massachusetts. Indeed, almost every summer between 1815 and 1823 someone reported seeing a sea serpent along the New England coast.

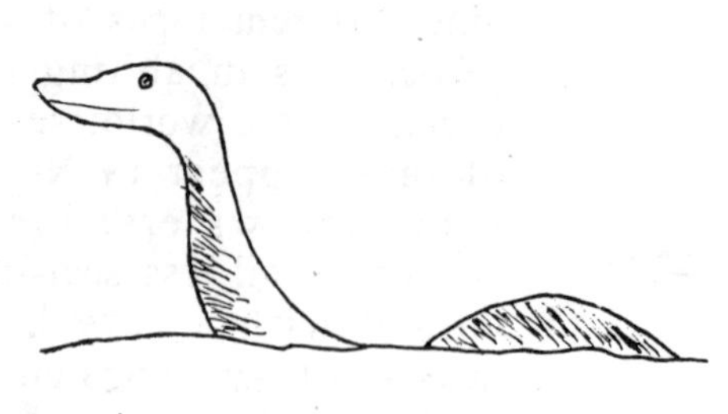

Super-eel

The Massachusetts Sea Serpent

On August 6, 1817 two women and several fishermen saw a huge monster sail into the harbor of Cape Ann, just north of Gloucester. Numerous fishermen also watched the event.

Four days later the monster appeared in the entrance to Gloucester Harbor, near Ten Pound Island. Seaman Amos Story watched the creature from the shore for an hour and a half. When he went to report the incident he learned there had been several other witnesses.

The news of the monster's arrival caused much excitement among coastal residents. Over the following weeks literally hundreds of people visited the slopes of Gloucester Harbor and many a respectable citizen laid his reputation on the line and swore under oath that he had personally seen the monster.

On August 14, four armed boats approached the monster and Matthew Gaffney, a ship's carpenter, fired at it from almost point-blank range. The musket ball struck its target in the head, but apparently had no effect.

On August 19, The Linnaean Society of New England set up a Committee of Inquiry to look into the matter. The committee comprised a judge, a physician and a naturalist. In a relatively short time they had collected a sizeable number of sworn statements from witnesses. The following excerpt is from an affidavit signed by Matthew Gaffney, the ship's carpenter:

> I, MATTHEW GAFFNEY, of Gloucester in the County of Essex...on the fourteenth day of August, A.D. 1817, between the hours of four and five o'clock in the afternoon...saw a strange marine animal, resembling a serpent, in the harbor in said Gloucester. I was in a boat and was within thirty feet (9 m) of him. His head appeared full as large as a four gallon keg, his body as large as a barrel, and his length that I saw, I should judge forty feet (12 m), at least. The top of his head was of a dark color, and the under part of his head appeared nearly white, as did several feet (a meter or so) of his belly...I supposed and do believe that the whole of his belly was nearly white. I fired at him...and think I must have hit him. He turned towards us immediately after I had fired, and I thought he was coming at us; but he sank down and went directly under our boat, and made his appearance at about one hundred yards (91.4 m) from where he sank. He did not turn down like a fish, but appeared to settle directly down, like a rock. My gun carries a ball of eighteen to the pound; and I suppose there is no person in town more accustomed to shooting than I am. I have seen the animal at several other times, but never had so good a view of him as on this day. His motion was vertical, like a caterpillar.

Gaffney swore on the bible as to the truth of his statement.

Considerable evidence collected from witnesses showed that descriptions matched remarkably with only minor differences. In effect, they saw an enormous smooth-skinned snake; dark brown, almost black in color, sometimes looking bluish except underneath its body, which was rather whitish. It moved very rapidly with a vertical flexing of its body. It was approximately sixty-five feet (20 m) long and usually appeared above the surface as a series of little humps—possibly as many as ten.

Convinced that the sea serpent was real, the authorities offered a reward for its capture. Traps, nets and hooks were set for it and an appeal went out to the Nantucket whalers. But the monster eluded them all and quietly slipped away to more hospitable waters.

About a month later, two little boys playing beside Loblolly Cove, near Cape Ann, found a small black snake. Along its back were strange humps. A farmer named Beach took the snake to the Linnaean Society with the suggestion that it was a young sea serpent, which had hatched from an egg laid by the monster during its recent visit. Surprisingly, they accepted his suggestion and declared it to be the "Atlantic humped snake" with the scientific name *Scoliophis atlanticus*.

When the Linnaean Society's detailed report came before the eyes of zoologists and scientists in Europe, the error was quickly seen. The "young sea serpent" was simply an ordinary black snake with a deformed spine—giving the impression of humps along its back!

Scorn was immediately thrown upon The Linnaean Society. The fact did remain, however, that the sea serpent was a reality—unless most of the population of Gloucester were hallucinating.

Like salt to the Linnaean Society's wound, the sea serpent returned to Gloucester Harbor again in July, 1818. On July 28, a boat loaded with fishermen armed with muskets fired seven or eight salvos at the creature, but it just ignored them and continued playing on the surface.

Next day, a captain and several whalers fired a harpoon and struck the monster a glancing blow. The result was that it promptly took refuge below the surface and in so doing, nearly capsized the boats! The attackers began to have second thoughts about the whole affair.

Every summer for the next ten years the sea serpent appeared along the east coast; its most favored spot being Massachusetts Bay. Successive sightings added new features to its picture; there was a horn or fin near its head and it had a mane.

Every time it raised its head from the water it made a noise like steam escaping from a boiler.

Over the years, local residents developed a fondness for the monster and his quiet unassuming ways. And when, in the summer of 1827, Captain David Thurlo, Jr., of the schooner *Lydia*, claimed he had harpooned a sea serpent off the coast of Maine, fans were very fearful for their lonely monster. Happily, their fears were unfounded as the sea serpent's dark elongated form was shortly seen once more along the east coast.

In recent years the Massachusetts sea serpent has not been seen again. Perhaps he took himself off to more friendly waters, or merely needed a change of scene.

Chessie, of Chesapeake Bay

Chesapeake Bay is a long, narrow arm of the Atlantic Ocean that runs north from the coast of Virginia and divides Maryland into two parts. Rivers flowing into the inlet include the great Potomac. The largest of the cities nearby are Washington, D.C. and Baltimore.

There have been intermittent reports of strange creatures in the area of Chesapeake Bay since the seventeenth century. But they did not receive much attention until 1965, when Pam Peters spotted a monster in the Hillsmere section of South River, near Annapolis. Descriptions usually concern a thirty to forty foot (9 to 12 m) long, snake-like creature reminiscent of the giant anaconda, dark in color, with a football sized head.

The serpent really came to the public's attention in 1977, when it received its name in an article in the *Richmond Times Dispatch*. In July, 1977 Gregg Hupka photographed the creature at the mouth of the Potomac River, but the photo was fuzzy and inconclusive.

There were many more encounters with Chessie over the next few years. Favorite spots for sightings were Love Point on Kent Island, the mouth of the Potomac River and Eastern Bay. Sightings usually took place between May and September.

Bill Burton, sports-fisherman and outdoor editor of the *Baltimore Evening Sun*—who has written extensively about sea serpent reports, suggests there might be a connection between the arrival of large bluefin in the area and Chessie sightings.

There were at least seventy-eight sightings prior to 1984,

many of which were by trained observers—a coast guard, two airline pilots, a naval officer, an ex-CIA official and an FBI agent. The reports also suggested there may be more than one serpent.

In an article in the *Chesapeake Bay Magazine*, Burton describes the experiences of the Reese family of Mathews, Virginia. Their home is beside the East River—a tributary of the Mobjack Bay, above York River. Apparently, early one morning in March, Bill Reese saw a greenish brown, twenty foot (6.1 m) long "hump-backed thing" swim upriver and back again. Its body was twelve inches (30 cm) in diameter and it moved with vertical undulations. His wife had also seen the creature the previous year.

Reese saw a smaller version of the sea serpent two months later, near his dock at Lyles Cove.

> ...I got in my rowboat, grabbed the oars and started after it. It swam to the other side of the cove, would go underwater and then come up just like the bigger one. The head, with eyes like a snake, was held out of the water about as high as a fist from the elbow...if you make a fist and turn your hand away, that is the way it looked...

It seems that Chessie is not only confined to the water. In 1978, oyster fisherman James Dutton found tracks that appeared to have been made by an enormous snake-like animal. The tracks crossed a field and entered Nanjemoy Creek, a tributary of the Potomac River, in Charles County. Two fishermen who saw the creature in the creek were so frightened they abandoned their boat!

Dutton also said, when he came to live in the Chesapeake Bay area forty years previously, there were many old ships rotting in the water. Some of these would have traded the South American routes. Perhaps a South American reptile, such as the fresh-water-living anaconda, was on board. It could have escaped into the water-way and somehow managed to survive. Chessie certainly had the size and shape of the anaconda. If the creature were a pregnant female, that might account for the multiple sightings.

There were several multiple sightings during 1978. Mary Lewis saw four or five of the creatures near the mouth of the Potomac River on June 27 and Donald Kyker, a retired CIA employee, saw four on July 21 near Heathsville. He shot at one of them, which resulted in public outcry. A movement was promptly begun by the Enigma Project and others to ensure that Chessie remain unmolested.

Sightings continued through 1979 to 1981. A very exciting encounter occurred on May 31, 1982, near Love Point on Kent Island. Robert Frew, an executive with a laser company and his family were entertaining guests at his home on Memorial Day, when they saw Chessie in Chesapeake Bay. Robert Frew grabbed his video camera and taped the whole performance.

Karen Frew described what the tape revealed:

> It shows two or three head shots—the head and the first seven or eight feet. At one point it came up and the whole thing surfaced and that's when we saw all thirty-five feet of it. One of the most notable things is the absence of any clear-cut markings. It was just...very dark brown to black.

The tape caused quite a stir and was analyzed by an eminent body of scientists at the Smithsonian Institution, in Washington. It was also put through a process of computer enhancement in an attempt to get more detail. But sadly, the outcome was inconclusive; the consensus of the panel being that the object on tape was "Animate, but Unidentifiable."

"The usual explanations of partially submerged logs, a string of birds or marine animals and optical illusion, seem inappropriate for the dark, elongated animal," stated George Zug, chairman of the Department of Vertebrate Zoology and a member of the board of directors of the International Society of Cryptozoology.

Craig Phillips, of the Division of Hatcheries and Fishery Management Services, of the U.S. Fish and Wildlife Service and formerly Director of the National Aquarium, also attended the Washington meeting. In a letter to the *I.S.C. Newsletter*, Winter 1982 issue, he considered a variety of animals that might be explanations for the creature on the tape. Moray or conger eels could be ruled out as they normally stay below the surface. Sea snakes are confined to tropical regions. Anacondas and pythons are not usually found in salt water. They like shallow water above 70° F (21° C).

Phillips went on to say that in 1953 he had experimented in this regard with a tame anaconda. He released it several times into the surf at Crandon Park Beach, Key Biscayne, Florida, but each time the snake swam back to shore. He also dismissed the idea that the creature on tape might be an oarfish; the silvery, flattened body in no way resembled Chessie's rounded, tube-like body.

Unaffected by his film debut and the controversy surrounding it, Chessie continued to make tantalizing appearances in Chesapeake Bay. There were at least four sightings in 1984 and others into the 1990s.

In 1985, there was a movement to encourage more scientific inquiry into Chessie's possible existence and to consider protecting the creature if it were ever caught. But the resolution, placed before a Senate Committee on Economic and Environmental Affairs in Annapolis, was defeated. The committee would not even acknowledge the creature's "possible existence."

I don't imagine the monster cares either way, but a lot of his fans and cryptozoologists certainly do.

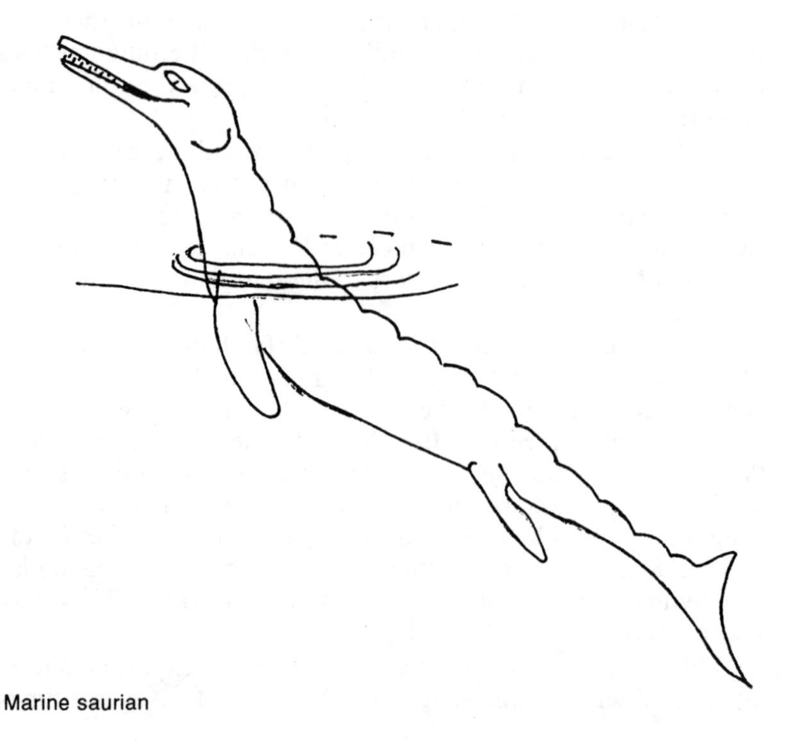

Marine saurian

"K. D." The Sea Serpent

13

Sea Serpents—The Gulf Of Mexico

Prior to 1840, sea serpent reports came mainly from the east coast. Then suddenly, in 1841, they began coming from the Gulf of Mexico. The Captain of the *Ville de Rochefort* came upon a sea serpent in the middle of the Gulf.

On July 14, there was another report from John Lloyd Stephens, the great American archaeologist who first discovered the Maya monuments in Yucatan:

> Towards evening we saw an enormous monster with straight black head, ten feet (3 m) out of the water, moving directly towards our ship. The Captain said it was not a whale. Another of the same kind appeared at the stern and we were really nervous; but were relieved by hearing them spout and seeing a column of water thrown into the air—at dark they were lying huge and motionless on the surface of the water.

There was not enough information in this report to determine what type of creatures these spouting monsters were, but there were similarities with the Massachusetts sea serpent.

On July 1, 1908 the *New York Times* told of a sighting by the steam ship *Livingstone*, of the Texas-Mexican Line, during a routine trip between Galveston, Texas and Frontera, Mexico. About fifty miles (80.5 km) out from Frontera, the monster appeared off the port bow. It was apparently sleeping on the surface.

Coming within sixty feet (18 m) of the creature, the ship

stood by for fifteen minutes while the entire crew watched it through the glasses. It was estimated to be "not less than 200 feet (61 m) long...about the diameter of a flour barrel in the center of the body, but not round." Its head was "approximately six feet (2 m) long by three feet (1 m) at the widest part. The color was dark brown and there were circles or rings near its tail."

Immediately upon docking in Frontera, the captain of the *Livingstone*, its crew and fifteen passengers, told their story to Charles N. Rickland, the U.S. Consular Agent. Afterwards they signed sworn affidavits as to the truth of their report.

On June 19, 1926 Miami *Daily News* printed a report of a "strange something" that was slowly rotting in an inlet at Sebastian, 150 miles (241.4 km) north of Miami. The "beast," which was discovered by turtle hunters, was reported by a fifteen-year-old boy named Claude.

Claude felt the creature was odd enough to deserve perpetuity. "Nothing like it has ever been seen or heard of in these waters," he wrote in a letter to the Daily News. "It is about fifteen feet (4.6 m) long and eight feet high (2.4 m)," and must have weighed around two tons. The head was like a camel's, there were no teeth and the body was covered by "very thick, coarse, grayish hair five or six inches (about 15 cm) long." There was "a hairy something resembling a tail."

Claude and his companions tried to cut into the carcass with a sharp knife and, after considerable effort, finally cut through the hair and into the thick hide. But they could not penetrate to the flesh. The boy was concerned that tides would soon cover the carcass with sand to the detriment of zoological investigation.

The boy's concern for scientific discovery was admirable and his report should have been taken seriously. However, since there appears to be no follow-up, it is likely his fears were well founded and the sea did reclaim the carcass before anything was done.

This sort of thing is very regrettable. As Heuvelmans says, wherever possible these obscure carcasses should be investigated by qualified people. "What if one day a real sea serpent, a big undiscovered animal, breathes its last on some beach and nobody but 'ignorant fishermen' stares at it, none but sea-birds dissect its flesh and only the waves take charge of its mighty bones?"

There was another sea serpent sighting in 1943, but it did not come to light until 1962, when Thomas Helm published his book *Monsters of the Deep*. Apparently, in 1943, Helm and his wife met a strange creature in St. Andrews Bay, off the northwest coast of Florida, in the Gulf of Mexico.

Helm had been invalided out of the U.S. Marines due to war injuries received at Pearl Harbor. He and his wife then became announcers at Panama City radio station. One afternoon towards the end of March, 1943 they were out in a small seventeen foot (5 m) catboat along the northwest coast of Florida. The sea was smooth and beautifully clear, as is often the case with sea serpent sightings.

Around 4:00 P.M. they were startled to see a huge animal with "a head about the size of a basketball on a neck which reached nearly four feet (1 m) out of the water," swimming directly towards their boat:

> The entire head and neck were covered with wet fur which lay close to the body and glistened in the afternoon sunlight. When it was almost beside our boat the head turned and looked squarely at us. My first thought was that we were seeing some kind of giant otter or seal...but this was not the face of an otter or seal. As a youngster I had run a trapline and knew mink and otter full well...The head of this creature, with the exception that there was no evidence of ears, was that of a monstrous cat. The face was fur covered and flat and the eyes were set in the front of the head.

Helm said the wet fur was a rich chocolate brown. The eyes were well-defined, round—about the size of a silver dollar and glistening black. Below the flattened black nose, there was a "moustache of stiff black hairs with a downward curve on each side."

For a brief moment the creature stared at the boat and then it turned away. Suddenly it looked back, as though in afterthought, and immediately dived beneath the surface. A swirl of foam and confusion near the boat indicated it had sounded close by.

Helm also said the creature did not in any way conform with reports he had read about unidentified sea creatures. But zoologist Bernard Heuvelmans thinks "the animal reminds one of several other descriptions and seems to be a smaller, more round-headed—in other words, a young specimen of the maned merhorse, or sea-foul."

At first Helm thought the creature was some sort of pinniped but then he realized this was not possible:

> ...Seals and sea lions have long pointed noses and the eyes are set on the sides of the head like those of a squirrel or rat. The creature my wife and I saw had eyes which were positioned near the front like those of a cat.

These features have also been mentioned by other witnesses of the maned sea serpent or merhorse.

Not all sea serpent sightings receive publicity at the time of happening. One incident, which occurred three years earlier, did not come to light until 1965, when the story was printed in *Fate* magazine.

It seems, in March 1962, while Edward Brian McCleary and three friends were skin diving from a rubber raft off the coast of Pensacola, Florida they became lost in the fog. Suddenly they heard a loud hissing noise and smelled a sickening odor. Out of the mist that surrounded them "a neck like a ten foot (3 m) pole" with bulbous head, green eyes and a serpentine body, struck at them. McCleary managed to escape to an old wreck of the sunken ship *Massachusetts*, upon which the four had intended to dive. But his friends had disappeared.

After the incident, McCleary told the story in detail to journalists, but it was never fully reported. Local newspapers made no mention of the sea serpent.

Perhaps the story sounded too fantastic to believe. But afterwards, when McCleary asked E. E. McGovern, the director of the search and rescue units, if he believed the story he replied: "The sea has a lot of secrets. There are a lot of things we don't know about. People don't believe these things because they're afraid to...I believe you. But there's not much else I can do."

In the winter of 1963, another weird carcass was washed up by the tide on the beach in Charlotte Harbor, on the west coast of Florida. Two teenage girls came upon the remains as they walked along the beach. The skull was about two and a half feet (76 cm) wide, with jellied eyes the size of tennis balls. Part of the jaw was spiked with teeth. The nostrils looked like blow holes but there were gills in the rooted sides. Behind the gills there were bony flippers with fringe. The exposed ribs were about the size of a calf's rib cage and dime-sized scales covered the skin.

Scientists at the State Marine Laboratory, at St. Petersburg, examined the remains, but could not identify the species. Dr. Eugenie Clark, marine biologist and oceanographer at the Cape Haze Marine Laboratory, in Sarasota, was equally baffled. The identity of this tantalizing gift from the sea still remains a mystery today.

The Serpent of Second Narrows, Burrard Inlet, Vancouver, B.C.

14

Sea Serpents—West Coast

The waters along the Pacific Coast near Vancouver, on the United States/Canadian border are rich in marine life, particularly salmon. This abundance supports a variety of seals, sea-lions, killer-whales and if reports are to be believed—sea serpents. Three varieties have been reported: the merhorse, a big-eyed long necked creature with huge body that lies just below the surface; the long necked sea serpent, with small eyes, a long neck and big body and a true serpentine animal.

Perhaps the most flamboyant of these is "Caddy," the cadborosaurus.

Caddy

According to a manuscript published in June 1973, entitled: *Observations of Large Unidentified Marine Animals in British Columbia and Adjacent Waters*, by Paul Le Blonde and John Sibert, the first encounter with a sea serpent near Vancouver was in 1905/6.

While working as a timber feller at Cracroft Island, Philip Welch of Port Alberni and a friend took a sixteen foot (4.8 m) rowboat to the north of Adams River in Johnstone Strait. They intended to fish for trout, but an enormous salmon run at the river mouth made this impossible.

By 9:00 A.M., they were ready to give up. Suddenly, a long, brown neck rose from the water about 200 yards (183 m) from the stern of the boat. Grabbing the oars, they rowed as fast as they could for shore. But the monster followed, moving faster than they could row. After a few anxious moments, the creature suddenly submerged and that was the last they saw of it.

Welch stated the creature's head was approximately six to eight feet (2 to 2.5 m) long and tapered from a twenty inch (51 cm) base to about eight to ten inches (20 to 25 cm) at the head end. The head was like a giraffe's, with two small five-inch (13 cm) horns and nostrils.

May 1922 produced another sighting near Pulteney Point lighthouse on Malcolm Island. J. Philips and C. G. Cook were on "stand-by" waiting for the lighthouse tender to arrive. Mr. Cook thought he saw the boat coming, but what he took to be the mast was the head and seven feet (2.1 m) of neck of a twenty-five foot (7.6 m) long strange sea creature. Its head moved as a snake's and had nostrils and large cow-like eyes. Its body was brown colored and appeared to be scaly.

Although they expressed relief when the creature passed by, they did say they were impressed with how gentle it appeared to be. This aspect has been noted repeatedly in sea serpent sightings and particularly where the merhorse is concerned.

In 1933, reports began coming in of sightings of a monster in the Strait of Georgia and Juan de Fuca Strait. Early reports came mainly from Canadian territory, between Vancouver Island and mainland British Columbia. But later reports showed the serpent also favored American waters—off Dungeness Spit and Whidbey Island, in Puget Sound. By its description, the serpent was again the typical merhorse.

The first report came from barrister and clerk to the British Columbia Legislature, Major W. H. Langley. Around 1:30 P.M., on Sunday October 1, 1933, Langley and his wife were out sailing in their yacht, *Dorothy*, near Discovery and Chatham Islands. Suddenly they heard a loud "grunt and a snort accompanied by a huge hiss," off the bow. They looked forward to see a large dark olive green colored creature near the edge of kelp beds off the Chatham Island shore. It had serrated markings along the top and sides.

Langley's report prompted one from F. W. Kemp, an officer of the Provincial Archives, who had seen the monster the year before. Kemp's report appeared in *The Victoria Daily Times*.

Apparently, on August 10, 1932 Kemp, his wife and son were on Chatham Island in the Juan de Fuca Strait when they

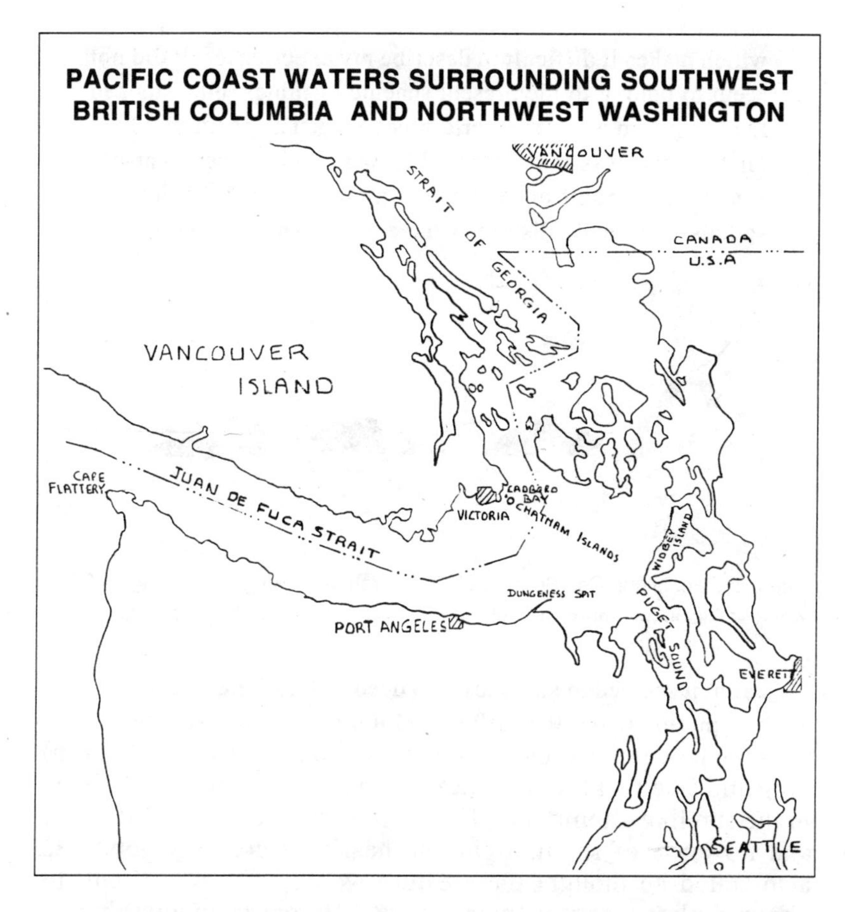

PACIFIC COAST WATERS SURROUNDING SOUTHWEST BRITISH COLUMBIA AND NORTHWEST WASHINGTON

noticed a mysterious something coming through the channel between Strong Tide Island and Chatham Island. The creature travelled with its head out of the water, at about four miles (6.4 km) per hour against the tide.

Swimming to the steep rocks of the island opposite, the creature put its head out of the water and up on the rocks, waving it from side to side as if taking its bearings.

> Then fold after fold of its body came to the surface. Towards the tail it appeared serrated with something moving flail-like at the extreme end. The movements were like those of a crocodile. Around the head appeared a sort of mane, which drifted round the body like kelp.
>
> The Thing's presence seemed to change the whole landscape,

which makes it difficult to describe my experiences. It did not seem to belong to the present scheme of things, but rather to the Long Ago when the world was young. The position it held on the rock was momentary. My wife and sixteen-year-old son ran to a point of land to get a clearer view. I think the sounds they made disturbed the animal. The sea being very

The Chatham Island Sea Serpent of 1932. From a drawing by drawing by F.W. Kemp. "This is my impression of the creature with its head on the rock..."

calm, it seemed to slip back into deep water; there was a great commotion under the surface and it disappeared like a flash.

Kemp estimated the creature was about eighty feet (24.4 m) long and five feet (1.5 m) thick. It was bluish green and "shone in the sun like aluminum." He felt it was capable of great speed and its sense of smell, sight and hearing were very good. He also stated he thought the creature would be very difficult to photograph, as its movements were different from anything he had ever heard of or seen.

Within a week of publishing Kemp's report, Archie Willis, news editor of *The Victoria Daily Times*, had received a dozen letters from other people who had also seen the monster. Willis felt such an important personage should have a name. He named the creature, Caddy, the cadborosaurus, after Cadboro Bay, where it was first seen.

When another slightly smaller creature appeared in the vicinity, it was believed to be Caddy's mate. She was given the name "Amy." She had a horselike head, no visible eyes or ears and, as befits a lady, no whiskers. Whether or not this specimen was a female of the species is not known. Indications are that it was simply Heuvelmans' long-necked variety.

Near the end of 1934, a rotting carcass was washed up on the beach at Henry Island, on the northwest coast of British Columbia. It was thirty feet (9.2 m) long and had reddish flesh and skin covered with hair mixed with quills. The head resembled a horse's and there were what looked like four fins or flippers attached to the spinal column. People wondered whether this might be one of Caddy's relatives.

The carcass was hauled up on the quay at Prince Rupert, where it was examined and photographed. Some of the photographs were printed in the *Illustrated London News*, on December 15, 1934.

After cursory examination, Dr. Neal Carter, Director of the Dominion Experimental Fisheries Station, commented the remains were those of a mammal and "in life it must have been slender and sinewy." But after further study at the Government Biological Station at Nanaimo, B.C., the director, Dr. Clements pronounced the remains as those of a basking shark.

By the 1940s Victoria residents had become very fond of their monsters and when, in the spring of 1943, Ernest Lee, skipper of a motor fishing boat announced he had rammed

Remains of a strange sea monster, it later turned out to be a basking shark.
Credit: Courtesy of the National Archives of Canada

Caddy twice off Vancouver Island, they considered it a very wicked deed. But their fears were unfounded, as the merhorse was shortly seen gamboling as usual in his favorite waters.

In April 1958, Caddy's family-oriented fans were delighted to hear a "baby" sea serpent had been seen undulating on the surface off Whidbey Island, in Puget Sound. The small specimen was purported to be approximately twelve feet (4 m) long and one foot (30 cm) thick. There were several witnesses, including the Reverend John Oosterhoff of Mount Vernon Presbyterian Church.

As one might expect, Caddy's fans decided the small specimen was one of his young. Both Caddy and Amy were seen frequently in the area at that time and in following years. The merhorse's flamboyant ways of playing on the surface while dozens of people watched from Victoria's beaches, endeared him to the hearts of local residents, who described him as "lovable and homely" and with "warm and kindly eyes."

Although he became known as Vancouver Island's resident sea serpent, the cadborosaurus has haunted both American and Canadian waters since at least 1928. Recent U.S. sightings were off Whidbey Island, in Puget Sound; off Dungeness Spit, in Juan de Fuca Strait; near north Pennock Island and in the icy waters of Security Bay, Alaska. These sightings and previous ones indicate the merhorse's territory extends at least from Juneau to Los Angeles.

Colossal Claude

South of Vancouver Island, the people of Oregon are convinced they also have an important personage in their waters. And naturally, they gave him a name: "Colossal Claude."

According to an article by Peter Cairns, in the September 24, 1967 edition of *The Sunday Oregonian*, the first encounter with Claude was in 1934. The crew of the Columbia River lightship and its supply tender, *Rose,* saw the forty foot (12.2 m) long creature swimming towards their boat. It had a neck about eight feet (2.4 m) long with snake-like head, a large bulky, round body and a large tail.

The crew wanted to lower a boat and investigate, but their captain felt caution was the best policy and gave the creature a wide berth.

In 1937, Claude appeared before the fishing boat, *VIV*. The skipper described the creature as about forty feet (12.2 m) long, hairy, tan-colored and with the head of an over-grown horse. It had a girth of four feet (1.2 m). Captain Chris Anderson of the *Argo* also accused the creature of raiding his fishing lines and swallowing a twenty pound (9 kg) halibut.

The length of the monster in both reports is the same but descriptions of a "snake-like head" and "the head of an overgrown horse," seem to be at odds. Otherwise, the descriptions are reminiscent of the merhorse.

The Stinson Beach Monster

Some of the most extra-ordinary sea serpent sightings in recent years on the west coast have occurred close to San Francisco. Like the coastal waters of British Columbia, the sea here is rich with marine life. All along the coast there are important breeding sites for harbor seals and Steller's and Californian sea-lions. Sea otters romp in the kelp beds. Elephant seals, almost at the point of extinction at the turn of the century, now number over 100,000. They come ashore at beaches from San Francisco to Baja, California. Blue whales feed off shore and the great white shark patrols, waiting for the frail or injured.

It comes as no surprise that a sea serpent would choose to frequent these waters. There have been a number of reported sightings since 1976.

The first indication of a monstrous presence in these waters appeared in the *Great Western Pacific Coastal Report* that year, in which Bolinas minister Tom D'Onofrio stated:

> On September 30, 1976 at 12 noon I experienced the most overwhelming event in my life. I was working on a carved dragon to use as a base for a table and couldn't complete the head. I felt compelled to go down to Agate Beach where I met a friend, Dick Borgstrom.
>
> Suddenly, 150 feet from shore, gamboling in an incoming wave, was this huge dragon, possibly sixty feet long and fifteen feet wide.
>
> The serpent seemed to be playing in the waves, threshing its tail. We were so overpowered by the sight we were rooted to the spot for about ten minutes. I literally felt as if I were in the presence of God. My life has changed since.

There was no newspaper publicity attached to this sighting until some seven years later, when there was another report of a monster in the vicinity.

At 2:30 P.M., on November 1, 1983, a road construction crew from the Californian Department of Transportation had an experience they will not soon forget. They were on Highway 1, a cliff-top road to the south of Stinson Beach in Marin County, north of San Francisco. The elevation was approximately 150 feet (45.7 m). Beneath them lay the sandy expanse of Stinson Beach and the vast waters of the Pacific.

In his book *There Are Giants in the Sea*, Michael Bright quotes from an interview he had with safety engineer and spokesperson for the crew, Marlene Martin:

> The flagman at the north end of the job-site hollered, "What's in the water?"
>
> We all looked out to sea but could see nothing, so the flagman, Matt Ratto, got his binoculars. I finally saw the wake and I said, "Oh my God, it's coming right at us, real fast."
>
> There was a large wake on the surface and the creature was submerged about a foot under water. At the base of the cliff, it lay motionless for about five seconds and we could look directly down and see it stretched out. I decided it must have been about 100 feet long, and like a big black hose about five feet in diameter. I didn't see the end of the tail.
>
> It then made a U-turn and raced back like a torpedo, out to sea. All of a sudden it thrust its head out of the water, its mouth went towards the sky and it thrashed about.
>
> Then it stopped, coiled itself up into three humps of the body and started again to whip about like an uncontrolled hosepipe. It did not swim sideways like a snake, but up and down.

Marlene said the creature resembled old drawings she had seen of dragons. She kept the binoculars focused on the head. The creature kept making coils, throwing its head about, splashing and opening its mouth. Its teeth were "peg-like" and even. She only noticed one of its eyes, which glowed a deep ruby color. It looked gigantic and ferocious. She did not see any fins or flippers and wondered how it could move so rapidly without them.

Although in some ways the creature resembled a snake-like

dinosaur, its movements were undulating and not from side to side like a snake's. She said she was stunned by the sight: "never in my wildest dreams could I have imagined a thing to be so huge and go so fast..."

Several sea-lions and many sea-birds were in the area when the creature arrived. After it left, all the wildlife disappeared for the rest of the day.

The incident was watched by all six crew members, who stood looking over the rail in disbelief. Truck driver, Steve Bjora, estimated the creature's speed to be fifty miles per hour (80 kph).

Fearing no one would believe their story, the construction crew decided to keep the experience to themselves. The story got out, however, when they were overheard talking about the day's events at a local pizza parlor. Other people then came forward, including D'Onofrio, to relate similar experiences.

There was a report, in February 1985, of another monster in San Francisco Bay. Enquiries showed there had been periodic sightings throughout the century of mysterious creatures off the Pacific Coast. A monster, approximately forty-five to seventy feet (14 to 21 m) long, seemed to have a penchant for Monterey Bay. Over a period of twenty years sightings had become so common local fishermen and residents had named the creature the "Old Man of Monterey Bay."

A sighting in 1946 off Cape San Martin, prompted further inquiries by the United Press. They learned the animal, dubbed "Bobo" by local residents, had appeared regularly over the previous ten years.

The San Clemente Monster

Between 1914 and 1919, there were intermittent reports of sightings of a strange creature between San Clemente Island and Santa Catalina, some fifty miles (80.5 km) south of Los Angeles. And as one might expect, the creature was subsequently dubbed the San Clemente monster.

Although Ralph Bandini, secretary of the Tuna Club, had heard about the monster, he had never actually seen it himself until one day in 1915. At the time, Bandini and his crewman, Percy Neal were out fishing for tunny ten miles (16 km) off

Catalina, in the San Clemente channel. Suddenly something huge and shining rose high and higher out of the water.

After that brief glimpse Bandini saw nothing more of the monster until September the following year. Bandini and a friend, Smith Warren were out in his power cruiser fishing for marlin. Around 8:00 A.M., about a mile (1.6 km) west of their camp at Mosquito Harbor, they suddenly saw something "dark and big heave up" in the water. Bandini describes this incident in his book, *Tight Lines*, published in 1932:

> I seized my glasses. What I saw brought me up straight! A great columnar neck and head...lifting a good ten feet (3 m). It must have been five or six feet (1.5 or 2 m) thick. Something that appeared to be a kind of mane of coarse hair, almost like a fine seaweed, hung dankly. But the eyes—those were what held me! Huge, seemingly bulging, round—at least a foot (30 cm) in diameter!
>
> We swung toward it...Then, even as I watched through the glasses, the Thing sank. There was no swirl, no fuss...just a leisurely, majestic sinking...

Bandini estimated the creature was bigger than the largest whale. It did not rise with the swell as a whale would have done, but remained motionless as a rock while the waves broke upon it.

He saw it on two other occasions in the same waters, but never at such close range. He was most impressed by its eyes, which were enormous, dull and indifferent like those of a corpse. These features, and the columnar neck, are significant of the merhorse type sea serpent.

Another description, given by George Farnsworth, president of the Tuna Club and the chief Californian sport-fishing-boat owner, appeared in *The History of the Tuna Club*:

> Its eyes were twelve inches (30 cm) in diameter, not set on the side like an ordinary fish, but more central. It had a big mane of hair about two feet long (61 cm)...It was some kind of mammal, for it could not have been standing so long unless it was.

In all, a large number of reputable witnesses had seen the San Clemente monster on various occasions, among them ex-president of the club, millionaire George C. Thomas III; Jimmy

Jump, holder of many fishing records; and Joe Coaxe, maker of the famous "Coxe" reel and expert on deep-sea fishing.

J. Charles Davis II, author of many technical fishing books, collected reports from witnesses, interviewing them separately. He wanted none to know of his interview with any of the others. "It was almost as though a recording had been made and each man played the same record," Davis stated. Each witness described a long columnous neck, big eyes like plates and a mane resembling seaweed.

For many sea serpent believers, this is the strength of the argument for the creature's existence—the repetitious nature of reports and the fact that, quite unknown to each other, witnesses are all describing the same creature. It is a persuasive aspect, but unfortunately, does not prove the case.

December 1950 saw the San Clemente monster in the news again. Miss Opal Lambert, who was working in the post office at Summerland, happened to look up from the Christmas cards she was stamping. Through the window she saw the merhorse swimming in circles about 200 yards (183 m) off shore. Head held four feet (1.2 m) above the surface, he continued amusing himself for some ten minutes before finally disappearing.

After that promising display, no more was seen of the monster until one very hot day in June 1953. Professional fisherman, Sam Randazzo and his crew of eight, were fishing in the outer Santa Barbara Channel when the monster suddenly popped up beside their boat. They immediately reported the incident to the coast guards:

> We saw the Thing, estimated that it had a neck ten feet (3.5 m) long and between five and six feet (1 to 2 m) thick. Its eyes were cone shaped, protruding, and about a foot (30 cm) in diameter.

Fate magazine also reported another sighting off Newport Beach in October, 1954 by Barney Armstrong, skipper of the *Sea-fern*.

Over the years, the San Clemente monster has been reported all along the Pacific Coast, from Monterey to Ensenada; its most favorite haunt being the San Clemente Channel and fairly close to the island after which it was named. And as seems to be the way with the merhorse (whom I suspect is a bit of a show-off), much of his time was spent playing on the surface to the amazement of onlookers.

The Sea-Hag of the Gulf of Georgia, British Columbia.

15

The Scene, In Review

Spring 1969, was an exciting time for monster enthusiasts in general. On April 15, the *M. V. Mylark*, a sixty-five foot (20 m) shrimp boat out of Kodiak, Alaska, was dragging for shrimp off Raspberry Island, in the Shelikoff Strait. The vessel was equipped with an echogram—a very sensitive sonar detection device, known as "Simrad." By emitting high-frequency sonic bursts, this Norwegian invention reproduces a graphic profile of the sea bottom and all sub-surface outlines on a continuous roll of graph paper. This device is also used by commercial fishermen in locating large schools of fish.

Although the *Mylark* did not expect to find anything of interest in that particular area, her Simrad was left operating. Suddenly, to the operator's astonishment, Simrad produced the clear outline of a huge 200 foot (61 m) long monster! It had a snub-nosed head attached to a long slender neck, two pairs of flippers and a long, tapering tail. It was, in effect, the perfect silhouette of the prehistoric plesiosaur.

Here, at last it seemed, was proof—sea serpents did exist! Professor A. C. Oudemans had believed this nearly eighty years ago. This also was the opinion of other eminent researchers, scientists and zoologists, such as British Commander/author R. T. Gould and Ivan T. Sanderson.

But much to the disappointment of enthusiasts, the world found Simrad's echogram hard to believe. Inevitably, the question was asked: had someone tampered with the strip in the machine? According to experts this was most unlikely. "Simrad is one of the most successful and widely used of all echo sounders. An outline recorded by this instrument is far more accurate and reliable than anything captured by the eye of a camera."

Zoologist Ivan Sanderson took the investigation in hand. He questioned the operator and studied the strip at length. In his opinion and that of fourteen other experts—attorneys, geographers, oceanologists and biologists, the strip was genuine. It was "concrete proof" (*Argosy* magazine, July 1970) that "somewhere in the icy waters off the southern coast of Alaska, there's at least one monstrous marine longneck swimming around—and who knows how many more?"

In November the following year, the hopes of sea serpent enthusiasts were again raised to dizzying heights. A rotting carcass was discovered shortly after high tide on a Mann Hill beach at Scituate, Massachusetts. Approximately thirty feet (9.1 m) long and estimated to weigh fifteen to twenty tons, the carcass was like nothing seen before.

Witnesses found it hard to describe. One person said it looked like a camel without legs. There appeared to be a long thin neck, a small head, thickish body with flippers and a long tail—the very features attributed to the sea serpent!

Biologists sped to the scene. Scores of eager sightseers carried away pieces of the carcass as souvenirs. But all hopes plummeted when experts identified the decaying remains as those of a basking shark. Their souvenirs suddenly worthless, scavengers quickly buried them and the whole disappointing affair.

With all the strange carcasses given up by the sea over the centuries and left to languish as "unknown species," one wonders whether science will ever be able to positively identify one as that of a sea serpent. Perhaps, as Ivan Sanderson suggests in his book, *Follow the Whale*, the term itself is too misleading. "Sea serpent" could really encompass many sea creatures who, due to some unknown factor in their environment had grown greatly beyond their norm.

"There presumably is no limit to size in oceans," Sanderson writes, "but there probably are limitations on form as one moves up in length. If there are 200 foot (61 m) creatures as yet undiscovered in the sea, they will probably prove to be greatly elongated or serpentine in form, whether they be seals, whales, or other mammals, or even fish."

Perhaps the most famous carcass ever to be found was the "Stronsa Beast," which was washed ashore in 1808 on one of the remote Orkney Islands, off the coast of Scotland. Many people believe the careless handling of this carcass resulted in the greatest loss ever to the science of marine zoology.

A fisherman named John Peace and his crew discovered the partly decayed marine animal in shallow water near a point known as Rothiesholm Head, on the island of Stronsay. It measured at least fifty-five feet (16.8 m) in length and again, much resembled the prehistoric plesiosaur. It had appendages that might once have been arms or flippers. A row of spines, each measuring approximately ten to fourteen inches (25 to 35 cm) long, ran down the middle of what appeared to be the back. John Peace broke off several of these spines and bones and placed them in the boat for future study.

For ten days winds and currents played fast and loose with the huge carcass. Finally it was driven ashore where Peace examined it again. The head was surprisingly small compared with the rest of the body and the bottom jaw was missing. The neck was slender and approximately fifteen feet (4.6 m) long. The circumference of the main part of the body was almost twelve feet (3.7 m). The long tapering tail appeared broken off beyond the halfway mark.

Realizing the scientific value of the carcass, Peace removed the head, one of the arm-like appendages and a piece of the hide which was still intact. A local artist made detailed sketches of the carcass.

Dozens of people viewed the remains, including many fishermen familiar with whales, basking sharks and other large creatures of the sea. They knew that the head and fore-parts of whales and sharks contain the greatest amount of overall bulk and that the mouth of both is enormous. The salvaged skull was slightly larger than that of a horse.

Peace sent the skull, some of the bones and sketches to various museums in Britain. It certainly looked as though the carcass of a true sea monster had at last been found. A year or so later, after much study of the bones, related material and masses of written reports, the verdict came. The dead animal was simply that old masquerader, the basking shark!

The controversy surrounding this verdict, however, continued for years, and there are many students of zoology today who still maintain the carcass of the Stronsa Beast was that of a *true* sea monster.

A great deal of research was done on the "Animal of Stronsa," by British Lieutenant-Commander R. T. Gould and is contained in his book, *The Case of the Sea Serpent*.

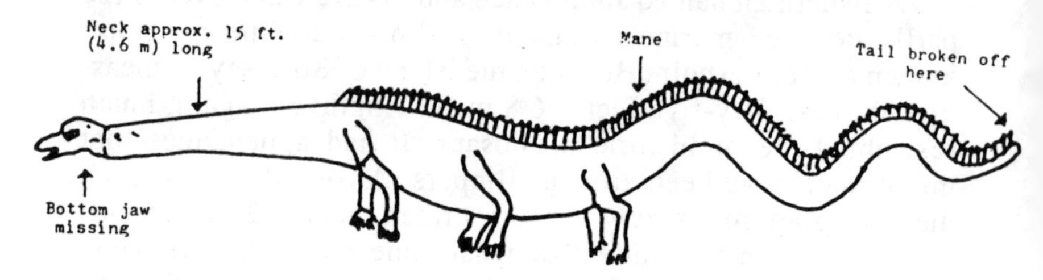

The Stronsa Beast of 1808. From the drawing by Mr. Petrie.

Another huge and equally famous carcass was washed ashore at Querqueville, France, in 1934. A skeleton of a "would-be" sea serpent that washed ashore along the New England coast, is encased in a museum at Harvard University; yet another, is preserved in California.

The difficulties involved in securing and positively identifying a carcass are considerable. The sea has a way of reclaiming its own before competent persons or zoologists can get to the scene and examine the carcass. The huge size prevents interested people from dragging the remains above the water line and out of reach of heavy seas. If the creature died at sea, its enormous weight and colossal size would take it quickly to the bottom, where scavengers would soon consume all trace.

Other difficulties are evident in the following story, from Tokyo, Japan, which appeared in the *Washington Post*, July 1977. While trawling off Christchurch, New Zealand, on April 10, Japanese fishermen snared in their nets at a depth of 1,000 feet, a mysterious two ton corpse of a plesiosaur-type reptile. When it was hoisted to the surface, the red and white carcass was estimated to be thirty-two feet (10 m) long. It had four flippers, a long neck and tail:

> The overpowering stench and the fatty liquids oozing onto the deck, overcame the fishermen's curiosity over the animal which neither they, nor any of the natural scientists since consulted, had ever seen before.

Fearful that the remains would spoil their cargo of eel-like whip-tail fish, the fishermen measured and photographed the carcass and then dumped it back in the ocean.

When the news of the dumping came to the ears of Japanese paleontologists, they were horrified. Such a thing should never have happened, they said. The skeleton should have been preserved for study.

Working with color photographs and sketches made by Michihiko Yano, a fishery company executive who happened to be on the trawler, they believed the remains were those of a species of plesiosaur known to have lived in the seas off eastern Australia. "It's not a fish, whale or any other mammal," said Professor Yoshinori Imaizumi, director general of animal research at the National Science Museum. "It's a reptile and the sketch looks very like a plesiosaur." The discovery seemed to indicate that these animals are not extinct after all.

Excited scientists, anxious to recover another of the creatures, instructed all fishing boats in the area to keep a watch for the original carcass and for live specimens. But fate was against them. The carcass was never recovered and no live specimens were seen.

The photographs have since been studied by a Scottish zoologist who has suggested the mysterious carcass might have been a Hooker's sea lion. But we shall never know for certain. Like all the other mysterious gifts from the sea, this one will forever remain a question.

The search for North America's monsters has led through forest and mountain wildernesses, hot deserts, deep cold lakes and the mysterious sea. We have examined reports of hundreds of witnesses who claim actually to have seen these creatures.

Witnesses have included ministers, lawyers, policemen, justices of the peace and government officials whose veracity is beyond question. Many witnesses have signed sworn testimonies as to the truth of their reports.

The sheer volume and repetitious nature of reports would seem proof enough that monstrous creatures, as yet unknown to science, do exist in the world today. Persuasive though this argument is, unfortunately, it is not acceptable. Science is based on fact not hearsay, or mere verbal reports. Only a specimen, carcass or some other positively identifiable remains will secure places for these strange phenomena in the annals of zoology.

The situation is not quite as bleak as it seems, however. A small group of biologists who form the International Society of Cryptozoology, are working to find species of land or sea animals hitherto completely unknown or to rediscover living examples of animals thought to have become extinct. Perhaps this dedicated group might be instrumental in bringing about acceptance of at least some of North America's "hidden" monsters.

Where the sasquatch, or Bigfoot is concerned, time itself might solve the mystery, as nowhere on earth remains inaccessible to man for long. With all the hunters—professional and amateur, combing the forests and wilderness areas of the Pacific Northwest and northern California, armed with guns and considerable determination, sooner or later this wildman/beast will be brought in, dead or alive.

Unfortunately, it is the only way to settle the question once and for all. We can only hope his captors are humane in their approach and their good intentions are not swayed by greed—for surely, fame and fortune await the first person to bring Bigfoot in.

When this time comes, some very important questions will have to be answered. Is the creature, as some have suggested, a kind of "half-man/half-beast," or is it simply a large animal whose roots lie in the age of dinosaurs—a creature who, because of his choice of remote environment has somehow managed to survive until relatively recently without discovery? It should be an interesting discussion.

Where lake monsters are concerned, irrefutable evidence could be obtained on film, if not by capture. Scientific expeditions are continually being undertaken not only in Scotland's Loch Ness, but in other lakes around the world.

Flying monsters? Here again, the camera might be the best, or only means of providing the ultimate answer to the question of their existence.

Proving the sea serpent's existence, however, is much more complicated. The very nature of the sea—its colossal size and depth, help keep its secrets hidden.

It is not likely a scientific search for sea monsters will ever be launched. The costs of such a venture would be too great. Unlike the situation with Bigfoot and the occasional lake monster, we don't have hunters setting out to sea in search of sea monsters or sea serpents. The job is such an enormous, costly and dangerous one, that it simply cannot be undertaken by ordinary people, or science, for that matter. Scientific sea probes undertaken so far are but a drop in the bucket compared with what would be required for a true sea monster search.

Information could come to us, however, in other ways. For instance, in November 1976 a completely unknown species of shark was accidentally brought to the surface from a depth of 500 feet (152 m) in a parachute anchor lowered by a U. S. Navy research vessel, off Hawaii. Almost fifteen feet (4.6 m) long, Megamouth—so named for its large mouth, was a plankton feeder and represents a completely new species, new genus and new family. Although its teeth are very small, there are 236 rows of them; they are structurally similar to those of the great white shark.

Unlike most sharks, which are fast and aggressive, Megamouth is quite sluggish and merely opens its large mouth to filter-feed on small organisms. The whale shark and the basking shark are the only other sharks which feed in this way.

According to Dr. Taylor, Director of the Waikiki Aquarium, in a 1983 interview with *Waikiki Beach Press*, "The discovery of megamouth...reaffirms science's suspicions that there are still all kinds of things—living in our oceans that we still don't know about. And that's very exciting."

Little by little we are learning more about the deep sea and the mysterious creatures that live in it. Scientists in the *Sealab* have glimpsed huge unknown creatures in the deep water. On two occasions during her thirty day underwater mission to explore 1,200 miles (1,931 km) of the Gulf Stream, the research submarine *Ben Franklin* spotted marine creatures ten times their normal size. The six men aboard reported to their surface support ship via sonar telephone that they had seen "eight blackfish thirty feet (9.1 m) long." According to Walter Muench, director of the *Ben Franklin* mission for the Grumman Aerospace Corporation, blackfish normally reach a maximum size of approximately three feet (1 m).

Mysteries of the deep are also constantly being revealed by the television and movie cameras of deep sea commercial engineering concerns. In 1962, to the astonishment of Californian oil drillers, their television cameras turned up photographs of what has been dubbed America's first *authentic* sea monster. The Shell Oil Company had started drilling operations two miles (3 km) off the coast near Santa Barbara. Using a "robot" driller controlled via a closed circuit underwater television camera, Shell photographer Forrest Adrian was waiting to take motion pictures of the pending drilling operations. His camera was set up on a tripod pointed at the television screen.

Shortly after midnight, a huge sea serpent suddenly appeared on the screen. Hardly able to believe his eyes, Adrian triggered the movie camera—thus, recording "live" the creature as it spiralled slowly through the water, brilliantly lit by the rig's illumination system.

"Marvin," as the sea serpent was nicknamed, appears on the film as a tapered creature about fifteen to twenty feet (4.5 to 6 m) long. It has a definite, though primitive head, eyes, mouth and a long thin tail. There are barnacled ridges along the body which moves through the water corkscrew fashion.

During the remaining night, Marvin returned several times, apparently attracted by the lights of the underwater camera. He created much interest among biologists and the public in general. His true identity, however, still remains a mystery. Miss Olga Hartmann, who specialized in invertebrate zoology for the Allen Hancock Foundation at the University of Southern California, saw the film and guessed that Marvin might be a "ctenaphor," an acorn-shaped jellyfish with tentacles. But the Scripps Institute of Oceanography suggests instead that the creature belongs to the salpida, a genus of transparent creatures that group themselves in long daisy chains. Biologists at the University of Texas, in Austin, believe Marvin is an unknown creature whose type has not changed for many centuries—something left over from ages past.

Marvin is not the first confirmed snakelike sea monster to come along. Sixty years ago, a Danish research ship under the direction of Dr. Johannes Schmidt of the University of Copenhagen, was trawling for deep-sea specimens off the west coast of Africa. When the nets were lifted to the surface, they were found to contain six foot (1.8 m) larval eels. Since all eel larvae grow at maturity to a size that is at least eighteen times larger than the grub form, a fully grown adult from such a larva would measure over 100 feet (30 m) long!

Modern technology has provided man with the tools to

make great sea journeys and explorations. Jacques-Yves Cousteau's little bright yellow, two-man diving saucer, or "Souscoupe"—the first of the *Deepstar* family of submersibles can easily explore depths to 1,200 feet (366 m). The largest *Deepstar* can explore to 20,000 feet (6,100 m) and the U.S. Navy's bathyscaphe *Trieste*, to 36,000 feet (11,000 m).

The U.S. Navy's *Trieste II*.

Credit: Official Photograph U.S. Navy

On January 23, 1960 under the command of U.S. Navy Lieutenant Don Walsh and Dr. Jacques Piccard, the bathyscaphe descended to the maximum depth in the Challenger Deep, off the Mariana Islands in the Pacific. Man had finally descended deeper than he had climbed. Mount Everest is 29,002 feet (8,846 m). The men suffered no ill effects from their seven mile (11 km) dive into the unknown. They brought back photographs of this dark submarine world which showed proof that life could indeed survive such crushing depths. They saw a fish very much like the common flounder, frequently found in shallow water of Long Island Sound and the Gulf of Mexico, swimming about miles below the surface.

The *Trieste* is only one of the deep submergence vehicles

whose "eyes," or movie cameras and television screens will bring us proof that in the fathomless depths of the sea, giant saurians swim as easily as gold fish in an aquarium.

Future deep-sea probes might produce creatures that amaze science and convince even the most rigid sea monster or sea serpent skeptics. The tide goes in and out like a great slow pulse; who knows what strange and marvellous creatures move beneath it?

Tables of Sightings

Lake and River Monsters of America

New England and Middle Atlantic States

DATE	LOCATION	DESCRIPTION
1609+	Lake Champlain, Vermont	Plesiosaur-like monsters, 20 to 45 feet long
?	Chain Lakes, Maine	?
1855	Silver Lake, New York	Hoax
?	Spirit Lake, Genesee County, New York	?
?	Sysladobosis Lake, Maine	Snake-like creature, head like a dog's
Sept. 1887	Wolf Pond, Pennsylvania	Snake-like monster, 30 feet long
1890s	Lake George, New York, off Boulton Landing	Possible hoax
1929	Lake of the Woods, New York, near Redwood	Plesiosaur-like monster
1951	Black River, New York	Monster 15 feet long, dark brown, round, tapered body; fins; "Eyes like silver dollars."
May 1968	MooreLake, New Hampshire	"Eerie" monster and "glowing red lights"
Mar. 1969	Hudson River, City Island Bridge, Bronx, New York	Enormous creature, larger than a whale; slimy, black and gray
?	Twin Lakes, Connecticut	Monster

Southern States

Date	Location	Description
1850+	White River, Arkansas	Monster, "Whitey"
1897	Mud Lake, Arkansas	Captured monster, 16 feet long
1953	Lake Conway, Arkansas	"Strange finny monsters"
1965	Reynolds Lake, near La Grange, Kentucky	Monster, body big as a stovepipe, large beady eyes
?	St. John's River, Florida	"A very primitive type of dinosaur"
1972/3	Herrington Lake, Kentucky	Monster
May 1975	North Fork St. Lucie River, Florida	30 feet aquatic creature
June 1975	St. John's River, Florida	Monster, "Pinky"

Midwestern States

DATE	LOCATION	DESCRIPTION
1800s+	Indiana (pools, sinkholes, rivers)	Reptilian monsters, horns like buffalo
-1838	Lake Manitou Indiana	Monster, 60 feet long
1867	Lake Michigan, Illinois coast	Serpent, 40 to 50 feet long
1867, 1882	Rock Lake, Wisconsin	Monster
1878	Mississippi River	Snake-like monster, pelican bill
1879-1968	Lake DuQuoin, Illinois	Alligator-like monster
1881	Lake Erie, Sandusky Bay	Serpent, 25 feet long
-1883+	Lake Mendota, Wisconsin	"Bozho," S.S.
1886	Great Sandy Lake, Minnesota	Monster
1889	Devil's Lake, Wisconsin	Two S.S.; fighting on surface
1890s	Lake Pewaukee, Wisconsin	A huge green "thing"
1890s	Elkhart Lake, Wisconsin	"Waterdragon"
1890s	Lake Koshkong, Wisconsin	Very large water animal
1890s	Lake Geneva, Wisconsin	S.S.—a "prankful monster"
1890s	Delavan Lake, Wisconsin	A "prankful monster"
1890s	Lake La Belle, Wisconsin	An immense fish, and serpent
1890s	Lake Okauchee, Wisconsin	An immense fish, and serpent
1890s	Fowler Lake, Wisconsin	An immense fish, and serpent
1890s	Oconomowoc Lake, Wisconsin	Serpent
1890s	Lake Kegonsa, Wisconsin	A vengeful and destructive monster
1890s	Lake Wingra, Wisconsin	A very large snapping turtle

1890s	Wabash River, Indiana	Large creature; leonine skull, whiskers
1891	Red Cedar Lake, Wisconsin	Snake-like creature, 50 feet long
1891	Lake Ripley, Wisconsin	Snake-like creature
1892	Lake Minnetonka, Minnesota	Turtle-like monster, 30 feet long
1892	Lake Huron	Serpent, 60 to 75 feet long
1897	Lake Monona, Wisconsin	S.S. 20 feet long
1900s	Milwaukee River, Wisconsin	Serpentine animal
1900s	Lake Michigan	Serpent
1920s	Lake Waubesa, Wisconsin	Giant eel-like creature, 60 to 70 feet long
1922	Paint River, Michigan	S.S.
-1922	Alkali Lake, Nebraska	Alligator-like monster, 40 feet long
1923	Big Alkali Lake, Nebraska	Reptilian monster, horns on head
1934	Big Chapman Lake, near Warsaw, Indiana	Monster, head 2 feet wide, "cow-like" eyes
1934	Lake Campbell, South Dakota	Reptilian monster
1935	Lake-of-the-Ozarks, Missouri	Large animal with snake-like neck
1971	Big Pine Lake, Minnesota	"Oscar," possibly a giant sturgeon
1976	Lake Huron, near Cheboygan, Michigan	Serpent
1976	Leech Lake, Minnesota	Monstrous fish

Southwestern States

DATE	LOCATION	DESCRIPTION
1973	Lake Eufaula, Oklahoma	"Nessie-type" monster

Rocky Mountain States

DATE	LOCATION	DESCRIPTION
1863 +	Bear Lake, Utah	"Nessie-type" monster
1868	Snake River, Idaho	Serpent
1877	Great Salt Lake, Utah	Reptilian monster, 75 feet long
-1882	Utah Lake, Utah	Reptilian monster
1886	Arizona Desert	Flying monster
1892	Lake DeSmet, Wyoming	Flying monster and water monsters
1900s+	Waterton Lake, Montana	Longnecked S.S.
1930-1950	Payette Lake, Idaho	"Slimie Slim" alias "Sharlie"

?	North Fork, Salmon River, Idaho	Black snake, 10 feet long, 5 inch diameter
1944	Lake Pendre Oreille, Idaho	Large unknown creature
-1956	Walker Lake, Nevada	Serpent, 45 to 50 feet long
1963	Flathead Lake, Montana	S.S.
1970	Missouri River, Montana	Black "fish"—6 to 8 feet long, rammed fiberglass boat, denting body

Pacific Coast States

DATE	LOCATION	DESCRIPTION
1800s	Crater Lake, Washington	"Strange creatures"
1800s	Crescent Lake, Washington	Strange "thing"
1830-1886	Lake Elizabeth, California	Flying monster
1892	Lake Chelan, Washington	Flying monster
1947-1964	Lake Washington, Washington	"Madrona sea monster"
1975	Lafayette Lake, California	Alligator-like creature
?	Lake Folsam, California	Monster

Alaska

DATE	LOCATION	DESCRIPTION
?	Iliamna Lake	Monsters, 20 feet long
?	Crosswind Lake, near Glenallen	Similar to Iliamna monsters
?	Lake Minchumina, between Fairbanks and Mt. McKinley	Similar to Iliamna monsters
?	Walter Lake, east of Kotzebue	Similar to Iliamna monsters
?	Nonvianuk Lake	Similar to Iliamna monsters

Lake and River Monsters of Canada

Province of Nova Scotia

DATE	LOCATION	DESCRIPTION
?	Lake Ainslie, Cape Breton Island	Serpentine monster, probably eel-balls
?	Bras d'Or Lakes, Cape Breton Island	Similar serpentine monsters

Province of New Brunswick

DATE	LOCATION	DESCRIPTION
?	Dungarvon River, a tributary of the Miramichi River	"Dungarvon Whooper"
1872+	Lake Utopia, Charlotte County	"Old Ned," a sea serpent
early 1880s	Lake Kilarney	Coleman frog

Province of Québec

DATE	LOCATION	DESCRIPTION
?	Aylmer Lake, 150 miles east of Montréal	Serpent
1894, 1896	Black Lake	Serpent, 20 to 30 feet long
1880	Lac Deschenes, on the Ontario/Québec border, an enlargement of the Ottawa River	Serpent, dark green, with body the thickness of a telegraph pole
1816, 1983, 1987	Lac Memphremagog, in Quebec's Eastern Township	"Memphre," a long-necked, snake-like monster
late 1800s+	Mocking Lake	Unknown animal 12 to 18 feet long, brown or black with saw-tooth fin
1930+	Lake Pohenegamook, 280 miles north east of Quebec City, near the New Brunswick border	"Ponick," a dark grey creature, 40 feet long

Province of Ontario

DATE	LOCATION	DESCRIPTION
?	Berens Lake, north of Red Lake, Northern Ontario	A denizen resembling an alligator

1938	Georgian Bay, Wasaga Beach, 20 miles west of Barrie	Monster which churned the water like an ocean liner, uncanny speed
1946, 1948	Lake of Bays, south east of Huntsville, seen off the north point of Fairview Island and Bigwin Island	Typical freshwater long-neck two humps, jet black, considerable wake
1600s+	Muskrat Lake, 60 miles west of Ottawa	"Nessie" type monster, 24 feet long, goes by the name of "Mussie"
?	Lake Miminiska, Northern Ontario, on the Albani River, west of Fort Hope	Fish-like Serpent
1829+	Lake Ontario, near Kingston	Serpent, "Kingstie"
?	Mazinaw Lake, 30 miles south east of Bancroft	Possibly a huge sturgeon
1960s	Nith River, south east Ontario, New Hamburg	"The Thing"—a three-toed, lizard-like creature, weighing approximately 50 pounds
1800, 1963	Lake Simcoe, 40 miles north of Toronto	Serpent, "Igopogo" approximately 70 feet long
1881+	Lake Simcoe, near Barrie, Kempenfeldt Bay	"Kempenfeldt Kelly" serpent 30 to 70 feet long
?	Rouge River, east of Toronto	"Lenny the Lizard," a pint-sized monster, purple ears and blue tail, 14 inches long

Province of Manitoba

DATE	LOCATION	DESCRIPTION
1909+	Lake Manitoba, Lake Winnipegosis	"Manipogo" and others, two-humped greenish-black serpent

Province of Saskatchewan

DATE	LOCATION	DESCRIPTION
1964	Rowan's Ravine, 50 miles north east of Regina	Creature 30 feet long, "looked like egg-shaped groups attached together"
early 1900s+	Turtle Lake, 50 miles east of Lloydminster	Monster 25 feet long, fin and three humps, long neck
?	Cold Lake on the Saskatchewan/ Alberta border	"Kinsoo," a large hump-backed serpent

Province of Alberta

DATE	LOCATION	DESCRIPTION
?	Battle River, 77 miles south east of Edmonton	Serpent, 25 to 30 feet long
1970s+	Christina Lake, 150 miles north of Saddle Lake	Unidentified monster 16 feet long
1970+	Saddle Lake, 100 miles east of Edmonton	Serpent, 18 feet long, horse-like head and hairy torso
?	Lake McGregor (an irrigation reservoir) and Saskatchewan River	Unidentified monster

Province of British Columbia

DATE	LOCATION	DESCRIPTION
?	Cowichan Lake, Vancouver Island	Freshwater longnecks called "Tsinquaws"
?	Harrison Lake, 80 miles east of Vancouver	Monster
1934	Lake Kathlyn, formerly Chicken Lake, Bulkley Valley	Serpent
1860+	Okanagan Lake, on the Pacific slopes of the Rockies	"Naitaka", serpent black/dark green, series of humps, 40 to 60 feet
1942	Oyster River, near Campbell River	"Klato," monster
?	Shuswap Lake	"TA-ZUM-A" monster, probably of Indian legend
?	Lake Tagai, Prince George	Monster about 10 feet long black, moves with great speed, creating large waves

Bibliography

Bell, Major Horace. *On the Old West Coast.* New York: Grosset and Dunlap, 1930.

Brown, Charles E. "Sea Serpents." *Wisconsin Folklore Society.* 1942.

Byrne, Peter. *The Search For Bigfoot, Monster, Myth or Man.* Washington: Acropolis Books, 1975.

Carson, Rachel. *The Sea Around Us.* New York: Oxford, 1951.

Cocking, Clive. "The Magical, Mystical, Mythical Sasquatch." *Weekend Magazine*, (May 10, 1975).

Cousteau, Jacques Yves. *World Without Sun.* New York: Harper and Row, 1964.

Chicago Sunday Tribune. Editorial page. July 24, 1892.

Clark, Jerome and Coleman, Loren. "America's Lake Monsters." *Beyond Reality*, (March - April, 1975).

Clark, Jerome and Lucuis Farish. "America's Mysterious 'Loch Ness' Monster." *Saga Magazine*, (November 1974).

Closse, Ben. "Loch Ness Monster Lives in Lake Champlain." *Midnight*, (March 15, 1971).

Cohen, Daniel. "Sea Serpents: what they really are." *Science Digest*, (March 1971).

Colbert, Edwin H. *Dinosaurs, Their Discovery and Their World.* New York: E. P. Dutton and Company, Inc. 1961.

Collier's Encyclopedia. (1972 ed.) VIII, pp. 225 - 235.

"Deep in the Dark Waters of Lake Conway Lurks a Terrible Monster (or maybe not)." *Log Cabin Democrat*, Conway, Arkansas. June 28, 1974.

Dinsdale, Tim. *The Leviathans.* London: Routledge and K. Paul. 1966.

______________. *Monster Hunt.* Washington: Acropolis Books Limited, 1972.

"Diving Saucer." *The New Book of Knowledge.* New York: Grolier Inc., (1974 ed.) XIX, 18.

Downs, R. B. *The Bear Went Over the Mountain.* Detroit: Singing Tree Press, 1971.

Drake, Cisco. "Flying Phantom of the West." *Beyond Reality.*

Fife, Austin E. "The Bear Lake Monsters." *Utah Humanities Review*, April, 1948. Vol.II, 2

Garner, Betty Sanders. *Canada's Monsters.* Hamilton, Ontario: Potlatch Publications, 1976.

Gillett, Edward. *Locating the Iron Trail.* Boston: Christopher Publishing House, 1925.

Gould, Rupert T. *The Case for the Sea-Serpent.* London: Philip Allan and Co., Limited, 1930.

Green, John. *On the Track of the Sasquatch.* British Columbia. Cheam Publishing Limited. 1968.

Grumley, Michael. *There are Giants in the Earth.* New York: Doubleday and Co. Inc., 1974.

Halpin, Marjorie and Michael M. Ames. *Manlike Monsters on Trial.* Vancouver: The University of British Columbia, 1980.

Hamlyn, Paul. *Larousse Encyclopedia of Animal Life.* England: The Hamlyn Publishing Group Limited, 1971.

Helm, Thomas. *Monsters of the Deep.* New York: Dodd, Mead and Company, 1962.

Heuvelmans, Bernard. *In the Wake of the Sea-Serpents.* London: Rupert Hart-Davis, 1968.

____________________. *On The Track of Unknown Animals.* London: Rupert Hart-Davis, 1958.

Holiday, F. W. *The Great Orm of Loch Ness.* London: Faber and Faber, 1968.

Hunter, Don with Dahinden, Rene. *Sasquatch.* Toronto: McClelland and Stewart, 1973.

Idyll, C. P. *Exploring the Ocean World.* New York: Thomas Y. Crowell Company Incorporated, 1969.

ISC Newsletters, and Journals: *Cryptozoology*, International Society of Cryptozoology, Tuscon, Arizona.

Kovalik, Vladimir, and Nada. *The Ocean World.* New York: Holiday House, 1966.

Leach, Maria. *Funk and Wagnalls' Standard Dictionary of Folklore, Mythology and Legend. New York: 1950.*

Lochbiler, Peter R. "Monsters surface off Cheboygan and that's no fish story." The Detroit News, June 24, 1975.

Markfield, Alan. "Professor Says He's Seen a Prehistoric Creature Swimming in a Kentucky Lake." Enquirer, (November 12, 1972).

Napier, John. Bigfoot, The Yeti and Sasquatch in Myth and Reality. London: Jonathan Cape Limited, 1972.

"Newport, Arkansas. Home of the White River Monster." *Federal Writers Project Little Rock, Arkansas*. Sponsored by the Newport Chamber of Commerce, 1937.

"New Monster Haunting River Called 'Pinky'." *Arkansas Gazette*, June 12, 1975.

Oudemans, Dr. A.C. *The Great Sea-Serpent*. London: Luzac and Company, 1892.

Paust, Gil. "Monster Mystery Fish." *Sports Afield*. (January, 1959.)

Porter, Majorie L. "The Champlain Monster." *Vermont Life*. (1970.)

Quinn, Daniel. *Land and Sea Monsters*. Northbrook, Illinois: Hubbard Press, 1971.

"Rather Fishy." *Forrest City Times*, Arkansas. May 28, 1897.

Ridgway, John and Blyth Chay. *A Fighting Chance*. Philadelphia:

J. B. Lippincott, Company, 1967.

Salkin, Harold. "Mysterious Water Monsters of North America."

Sea Monsters, Spring, 1977.

Sanderson, Ivan T. "Alaska's Sea Monster." *Argosy*. (July 1970.)

__________________. "First Photos of 'Bigfoot', California's Legendary 'Abominable Snowman'." *Argosy*. (February, 1968.)

__________________. "More Evidence that Bigfoot Exists." *Argosy*, 1968.

__________________. "Monster Hunting." *Saga Magazine*, (January 196?)

__________________. "The Big Lake Monster Hunt." *Fate Magazine*, (November, 1964).

__________________. "The Five Weirdest Wonders of the World." ? (November, 1968)

Simmonds, A. J. "Utah's Mountain Lake Monsters." *True Frontier*, (January 1969).

Smalley, Donald. "The Logansport Telegraph and the Monster of the Indiana Lakes." *Indiana Magazine of History*, Vol.II 1946.

Steiger, Brad. "Watermonsters." *Male Magazine*, (March, 1971).

"Strange Fauna." *Idaho Lore, Federal Writers Program*, 1939.

Sweeney, James B. *A Pictorial History of Sea Monsters and Other Dangerous Marine Life. New York: Crown Publishers, 1972.*

Vermont Life. Spring 1962.

"White River Monster." Pursuit, (vol 4, 1971).

Wolkomir, Richard. "The Glowing 'Thing' in Moore Lake." Fate Magazine, (November, 1968).

Wylie, Kenneth. *Bigfoot, A Personal Inquiry Into a Phenomenon*. New York: The Viking Press, 1980.

Assorted unidentified newspaper clippings from the files of Tim Church and Pat Oleksiw.

Recommended Reading

Green, John. *Sasquatch, The Apes Among Us*. British Columbia, Canada: Cheam Publishing Limited and Hancock House Publishers Ltd., 1978.

Heuvelmans, Bernard. *In the Wake of the Sea-Serpents*. London: Rupert Hart-Davis, 1968.

For monster enthusiasts in general or those interested in the discovery of little known, or previously undiscovered animals, I recommend membership in the International Society of Cryptozoology, P.O. Box 43070, Tuscon, Arizona 85733, U.S.A.